# Contents

**Preface** ix

### SECTION I. TRIGONOMETRY

**1 Trigonometric ratios for angles of any magnitude** 1
Main points 1
Worked problems 5
Further problems 11

**2 The solution of triangles and their areas** 13
Main points 13
Worked problems 14
Further problems 21

**3 Trigonometric graphs and the combination of waveforms** 25
Main points 25
Worked problems 29
Further problems 37

**4 Trigonometric identities and the solution of equations** 39
Main points 39
Worked problems 40
Further problems 47

### SECTION II. MENSURATION

**5 Areas and volumes** 48
Main points 48
Worked problems 49
Further problems 56

**6 Irregular areas and volumes and mean values of waveforms** 58
Main points 58
Worked problems 60
Further problems 65

**7 Centroids of simple shapes** 67
Main points 67
Worked problems 68
Further problems 74

**8 Further areas and volumes** 78
Main points 78
Worked problems 79
Further problems 87

### SECTION III. STATISTICS

**9 Presentation of grouped data** 90
Main points 90
Worked problems 91
Further problems 95

**10 Measures of central tendency and dispersion** 99
Main points 99

    Worked problems   101
    Further problems   105

## 11 An introduction to normal distribution curves   107
    Main points   107
    Worked problems   108
    Further problems   114

## 12 Probability   115
    Main points   115
    Worked problems   116
    Further problems   121

## SECTION IV. GRAPHS

## 13 The straight line graph   124
    Main points   124
    Worked problems   126
    Further problems   130

## 14 Reduction of non-linear laws to linear form   133
    Main points   133
    Worked problems   134
    Further problems   141

## 15 Graphs with logarithmic scales   144
    Main points   144
    Worked problems   146
    Further problems   156

## 16 Graphical solution of equations   156
    Main points   156
    Worked problems   158
    Further problems   165

## SECTION V. ALGEBRA

## 17 The solution of equations   167
    Main points   167
    Worked problems   168
    Further problems   176

## 18 Exponential functions and Naperian logarithms   179
    Main points   179
    Worked problems   182
    Further problems   189

## 19 Boolean algebra and switching circuits   192
    Main points   192
    Worked problems   195
    Further problems   200

## SECTION VI. CALCULUS

## 20 Introduction to differentiation   204
    Main points   204
    Worked problems   207
    Further problems   210

**21 Introduction to integration** 212
  Main points   212
  Worked problems   214
  Further problems   220

**Index**   223

# Note to readers

*Checkbooks* are designed for students seeking technician or equivalent qualification through the courses of the Business and Technician Education Council (BTEC), the Scottish Technical Education Council, Australian Technical and Further Education Departments, East and West African Examinations Council and other comparable examining authorities in technical subjects.

*Checkbooks* use problems and worked examples to establish and exemplify the theory contained in technical syllabuses. *Checkbook* readers gain real understanding through seeing problems solved and through solving problems themselves. *Checkbooks* do not supplant fuller textbooks, but rather supplement them with an alternative emphasis and an ample provision of worked and unworked problems, essential data, short answer and multi-choice questions (with answers where possible).

Electrical Applications 2
Electrical and Electronic Principles 2
Electrical Principles 3
Electronics 2
Electronics 3
Engineering Science 2
Engineering Science 3
Mathematics 1

Mathematics 2
Mathematics 3
Mathematics 4
Microelectronic Systems 1
Microelectronic Systems N2
Microelectronic Systems 3
Science F
Workshop Processes and Materials 1

# Preface

This textbook of worked problems provides coverage of the Technician Education Council level 2 units in Mathematics. However, it can be regarded as a basic textbook in Mathematics for a much wider range of courses. Each topic considered in the text is presented in a way that assumes in the reader only the knowledge attained at TEC level 1 Mathematics or its equivalent (GCE O level or CSE I in Mathematics).

The aim of the book is to extend the fundamentals in Mathematics gained in the Level 1 unit or to consolidate the mathematical ability of those students entering Level 2 directly. The essential theme is one of awareness of mathematical concepts. Mathematics 2 provides a follow-up to the checkbook written for Mathematics 1.

This practical Mathematics book contains nearly 250 detailed worked problems, followed by some 400 further problems with answers.

The authors would like to express their appreciation for the friendly co-operation and helpful advice given to them by the publishers. Thanks are also due to Mrs Elaine Mayo for the excellent typing of the manuscript. Finally, the authors would like to add a word of thanks to their wives, Elizabeth and Juliet, for their continued patience, help and encouragement during the preparation of this book.

J O Bird
A J C May
Highbury College of Technology
Portsmouth

# 1 Trigonometric ratios for angles of any magnitude

## A. MAIN POINTS CONCERNED WITH TRIGONOMETRIC RATIOS FOR ANGLES OF ANY MAGNITUDE

1  **Trigonometry** is concerned with the measurements of the sides and angles of triangles and their relationships with each other.

   **Trigonometric ratios of acute angles**

2  With reference to the right angled triangle shown in *Fig 1*:

   (i) sine $\theta = \dfrac{\text{opposite side}}{\text{hypotenuse}}$, i.e. $\sin \theta = \dfrac{b}{c}$

   (ii) cosine $\theta = \dfrac{\text{adjacent side}}{\text{hypotenuse}}$, i.e. $\cos \theta = \dfrac{a}{c}$

   (iii) tangent $\theta = \dfrac{\text{opposite side}}{\text{adjacent side}}$, i.e. $\tan \theta = \dfrac{b}{a}$

   (iv) secant $\theta = \dfrac{\text{hypotenuse}}{\text{adjacent side}}$, i.e. $\sec \theta = \dfrac{c}{a}$

   (v) cosecant $\theta = \dfrac{\text{hypotenuse}}{\text{opposite side}}$, i.e. $\operatorname{cosec} \theta = \dfrac{c}{b}$

   (vi) cotangent $\theta = \dfrac{\text{adjacent side}}{\text{opposite side}}$, i.e. $\cot \theta = \dfrac{a}{b}$

**Fig. 1**

3  From para. 2, (i) $\dfrac{\sin \theta}{\cos \theta} = \dfrac{\frac{b}{c}}{\frac{a}{c}} = \dfrac{b}{a} = \tan \theta$, i.e. $\tan \theta = \dfrac{\sin \theta}{\cos \theta}$

   (ii) $\dfrac{\cos \theta}{\sin \theta} = \dfrac{\frac{a}{c}}{\frac{b}{c}} = \dfrac{a}{b} = \cot \theta$, i.e. $\cot \theta = \dfrac{\cos \theta}{\sin \theta}$

   (iii) $\sec \theta = \dfrac{1}{\cos \theta}$

   (iv) $\operatorname{cosec} \theta = \dfrac{1}{\sin \theta}$ (Note: 's' and 'c' go together)

   (v) $\cot \theta = \dfrac{1}{\tan \theta}$

   Secants, cosecants and cotangents are called the **reciprocal ratios**.

4  **4-figure trigonometrical tables** provide values for each of the six trigonometric ratios from 0° to 90° correct to the nearest minute. Tables of secants, cosecants and cotangents are read in a similar manner to those used for sines, cosines and tangents.

(i) From **natural secant tables** the value of a secant is seen to increase from 1 at 0° to infinity (∞) at 90°.

For example,  sec 34°     = 1.2062
sec 34° 18′ = 1.2105
sec 34° 22′ = 1.2105+10 (from the mean difference column equivalent to 4′)
i.e. sec 34° 22′ = 1.2115.

Similarly, sec 21° 53′ = 1.0776 and sec 63° 10′ = 2.2155.

(Using a calculator containing only sine, cosine and tangent functions to find sec 34° 22′, the procedure is:

(a) Convert the minutes to degrees: $22/60 = 0.3666667$
(b) +34 gives 34.366667°
(c) cos 34.366667° = 0.825442
(d) Since $\sec \theta = \dfrac{1}{\cos \theta}$ then $\dfrac{1}{0.825442} = 1.2114722$,

i.e. sec 34° 22′ = 1.2115, correct to 4 decimal places.)

To find the angle whose secant is 1.4732 find in the table the nearest number **less than** 1.4732; in this case it is 1.4718 corresponding to 47° 12′. 1.4718 is 14 less than 1.4732 and 14 in the mean difference column is equivalent to 3′. Hence the angle whose secant is 1.4732 is 47° 12′+3′, i.e. 47° 15′. Hence arcsec 1.4732 = 47° 15′ ('arcsec $\theta$' is a short way of writing 'the angle whose secant is equal to $\theta$'). Similarly, arcsec 1.1240 = 27° 10′ and arcsec 2.3172 = 64° 26′. (Using a calculator to find arcsec 1.4732:
**Procedure**:
(a) $1/1.4732 = 0.6787945$; (b) arccos 0.6787945 = 47.250487°;
(c) −47 gives 0.250467; (d) 0.250467 × 60 = 15.02922′; i.e.
arcsec 1.4732 = 47° 15′, correct to the nearest minute.)

(ii) From **natural cosecant tables**, the value of a cosecant is seen to decrease from infinity (∞) at 0° to 1 at 90°.

For example,  cosec 22°     = 2.6695
cosec 22° 24′ = 2.6242
cosec 22° 27′ = 2.6242−55 (from the mean difference column equivalent to 3′)
i.e. cosec 22° 37′ = 2.6187

(*Note*: mean differences are **subtracted** for cosecant values.)
Similarly, cosec 41° 8′ = 1.5202 and cosec 77° 46′ = 1.0232. To find the angle whose cosecant is 1.8941, find in the table the nearest number **greater than** 1.8941; in this case it is 1.8977 which corresponds to 31° 48′. 1.8941 is 36 less than 1.8977 and 36 in the mean difference column corresponds to 4′. Hence arccosec 1.8977 = 31° 48′ + 4′, i.e. 31° 52′. Similarly, arccosec 2.9476 = 19° 50′ and arccosec 1.1248 = 62° 45′.

(iii) From **natural cotangent tables**, the value of a cotangent is seen to decrease from infinity (∞) at 0° to 1 at 45° and then to zero at 90°. Natural cotangent tables are read in a similar way to natural cosecant tables. For example, cot 21° 16′ = 2.5695 and cot 72° 49′ = 0.3093. (Note that the mean difference is subtracted since the cotangents are decreasing in value as the angles are increasing.) Also, arccot 1.7298 = 30° 2′ and arccot 0.4316 = 66° 39′.

### Angles of any magnitude

5 (i) *Fig 2* shows rectangular axes XX' and YY' intersecting at origin 0. As with graphical work, measurements made to the right and above 0 are positive whilst those to the left and downwards are negative. Let OA be free to rotate about 0. by convention, when OA moves anticlockwise angular measurement is considered positive, and vice versa.

(ii) Let OA be rotated anticlockwise so that $\theta_1$ is any angle in the first quadrant and let perpendicular AB be constructed to form the right-angled triangle OAB (see *Fig 3*). Since all three sides of the triangle are positive, all six trigonometric ratios are positive in the first quadrant. (*Note*: OA is always positive since it is the radius of a circle.)

(iii) Let OA be further rotated so that $\theta_2$ is any angle in the second quadrant and let AC be constructed to form the right-angled triangle OAC. Then:

$$\sin \theta_2 = \frac{+}{+} = +, \quad \cos \theta_2 = \frac{-}{+} = -, \quad \tan \theta_2 = \frac{+}{-} = -,$$

$$\operatorname{cosec} \theta_2 = \frac{+}{+} = +, \quad \sec \theta_2 = \frac{+}{-} = -, \quad \cot \theta_2 = \frac{-}{+} = -.$$

(iv) Let OA be further rotated so that $\theta_3$ is any angle in the third quadrant and let AD be constructed to form the right-angled triangle OAD. Then:

$$\sin \theta_3 = \frac{-}{+} = - \text{ (and hence cosec } \theta_3 \text{ is } -), \quad \cos \theta_3 = \frac{-}{+} = - \text{ (and hence}$$
$$\sec \theta_3 \text{ is } -), \quad \tan \theta_3 = \frac{-}{-} = + \text{ (and hence cot } \theta_3 \text{ is } +).$$

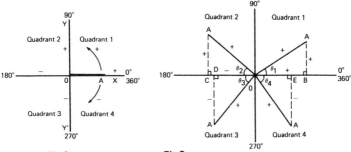

**Fig 2**  **Fig 3**

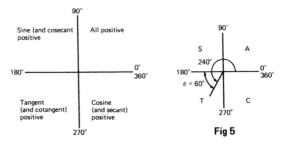

**Fig 4**  **Fig 5**

(v) Let OA be further rotated so that $\theta_4$ is any angle in the fourth quadrant and let AE be constructed to form the right angled triangle OAE. Then:
$\sin \theta_4 = \frac{-}{+} = -$ (and hence cosec $\theta_4$ is $-$), $\cos \theta_4 = \frac{+}{+} = +$ (and hence sec $\theta_4$ is $+$), $\tan \theta_4 = \frac{-}{+} = -$ (and hence cot $\theta_4$ is $-$).

(vi) The results obtained in (ii) to (v) are summarised in *Fig 4*. The letters underlined spell the word **CAST** when starting in the fourth quadrant and moving in an anticlockwise direction.

(vii) To evaluate sin 240°:
 (a) Sketch rectangular axes and mark on angles and the word CAST as shown in *Fig 5*.
 (b) Mark on the sketch an angle of 240°.
 (c) Determine angle $\theta$ (= 240° − 180° = 60°), which is always measured to the horizontal.
 (d) Use tables to evaluate sin 60° (= 0.8660).
 (e) Since T is shown in the third quadrant, then only tangent and cotangent are positive in this quadrant, i.e. sine is negative.
 Hence sin 240° = −sin 60° = −0.8660

## 6 Graphs of trigonometric functions

By drawing up tables of values from 0° to 360°, graphs of $y = \sin A$, $y = \cos A$ and $y = \tan A$ may be plotted. Tables, and a knowledge of angles of any magnitude, are needed (or alternatively, a calculator). Such tables, using 30° intervals, are shown below with the respective graphs shown in *Fig 6*.

(a) $y = \sin A$

| A | 0 | 30° | 60° | 90° | 120° | 150° | 180° | 210° | 240° | 270° | 300° | 330° | 360° |
|---|---|---|---|---|---|---|---|---|---|---|---|---|---|
| sin A | 0 | 0.50 | 0.866 | 1.00 | 0.866 | 0.50 | 0 | −0.50 | −0.866 | −1.00 | −0.866 | −0.50 | 0 |

(b) $y = \cos A$

| A | 0 | 30° | 60° | 90° | 120° | 150° | 180° | 210° | 240° | 270° | 300° | 330° | 360° |
|---|---|---|---|---|---|---|---|---|---|---|---|---|---|
| cos A | 1.00 | 0.866 | 0.50 | 0 | −0.50 | −0.866 | −1.00 | −0.866 | −0.50 | 0 | 0.50 | 0.866 | 1.00 |

(c) $y = \tan A$

| A | 0 | 30° | 60° | 90° | 120° | 150° | 180° | 210° | 240° | 270° | 300° | 330° | 360° |
|---|---|---|---|---|---|---|---|---|---|---|---|---|---|
| tan A | 0 | 0.577 | 1.732 | ∞ | −1.732 | −0.577 | 0 | 0.577 | 1.732 | ∞ | −1.732 | −0.577 | 0 |

From *Fig 6* it is seen that:
(i) Sine and cosine graphs oscillate between peak values of ± 1.
(ii) The cosine curve is the same shape as the sine curve but displaced by 90°.
(iii) In the first quadrant all the curves have positive values; in the second only sine is positive; in the third only tangent is positive; in the fourth only cosine is positive (exactly as summarised in *Fig 4*).
(iv) The sine and cosine curves are continuous and they repeat at intervals of 360°; the tangent curve appears to be discontinuous and repeats at intervals of 180°.

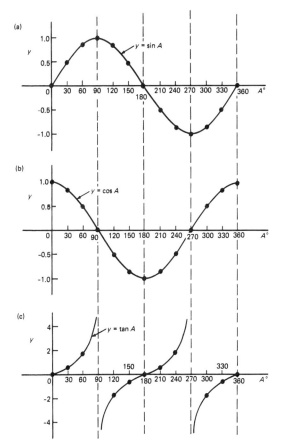

**Fig 6**

## B. WORKED PROBLEMS ON TRIGONOMETRIC RATIOS FOR ANGLES OF ANY MAGNITUDE

*Problem 1* Determine the values of the six trigonometric ratios for angle $\theta$ shown in the right-angled triangle ABC shown in *Fig 7*

By definition:   $\sin \theta = \dfrac{\text{opposite side}}{\text{hypotenuse}} = \dfrac{5}{13} = 0.3846$   Hence $\operatorname{cosec} \theta = \dfrac{13}{5} = 2.6000$

$\cos \theta = \dfrac{\text{adjacent side}}{\text{hypotenuse}} = \dfrac{12}{13} = 0.9231$   Hence $\sec \theta = \dfrac{13}{12} = 1.0833$

$\tan \theta = \dfrac{\text{opposite side}}{\text{adjacent side}} = \dfrac{5}{12} = 0.4167$   Hence $\cot \theta = \dfrac{12}{5} = 2.4000$

*Problem 2* If cos $X$ = 8/17 determine the values of the other five trigonometric ratios.

*Fig 8* shows a right-angled triangle. Since cos $X$ = 8/17, then XY = 8 units and XZ = 17 units.
Using Pythagoras' theorem: $17^2 = 8^2 + YZ^2$
from which $YZ = \sqrt{(17^2 - 8^2)} = 15$ units.
Thus $\sin X = \frac{15}{17}$, $\tan X = \frac{15}{8} = 1\frac{7}{8}$,
cosec $X = \frac{17}{15} = 1\frac{2}{15}$, sec $X = \frac{17}{8} = 2\frac{1}{8}$
and cot $X = \frac{8}{15}$.

**Fig 8**

*Problem 3* Use 4 figure tables to evaluate (a) sec 19° 28'; (b) cosec 49° 17'; (c) cot 36° 51'.

(a)  sec 19° 24'  = 1.0602
     sec 19° 28'  = 1.0602+4 (from the mean difference column
                             equivalent to 4')
i.e. **sec 19° 28'  = 1.0606**

(b)  cosec 49° 12'  = 1.3210
     cosec 49° 17'  = 1.3210−16 (from the mean difference column
                                 equivalent to 5')
i.e. **cosec 49° 17'  = 1.3194**

(c)  cot 36° 48'  = 1.3367
     cot 36° 51'  = 1.3367−25 (from the mean difference column
                               equivalent to 3')
i.e. **cot 36° 51'  = 1.3342**

*Problem 4* From 4 figure tables find the actue angles (a) arcsec 2.3161; (b) arccosec 1.1784; (c) arccot 2.1273.

(a) The nearest number less than 2.3161 in natural secant tables is 2.3144, corresponding to 64° 24'. The difference between 2.3144 and 2.3161 is 17. From the mean difference column 17 corresponds closest to 1'.
Hence arcsec 2.3161 = 64° 24'+1' = **64° 25'**.
(b) The nearest number greater than 1.1784 in natural cosecant tables is 1.1792 corresponding to 58° 0'. The difference between 1.1784 and 1.1792 is 8. From the mean difference column 8 corresponds to 4'.
Hence arccosec 1.1784 = 58° 0'+4' = **58° 4'**.
(c) The nearest number greater than 2.1273 in natural cotangent tables is 2.1348 corresponding to 25° 6'. The difference between 2.1273 and 2.1348 is 75. From the mean difference column 75 corresponds closest to 5'.
Hence arccot 2.1273 = 25° 6'+5' = **25° 11'**.

*Problem 5* If $\sin \theta = 0.625$ and $\cos \theta = 0.500$ determine, without using trigonometrical tables or calculators, the values of cosec $\theta$, sec $\theta$, tan $\theta$ and cot $\theta$.

From para. 3, $\operatorname{cosec} \theta = \dfrac{1}{\sin \theta} = \dfrac{1}{0.625} = \mathbf{1.60}$

$\sec \theta = \dfrac{1}{\cos \theta} = \dfrac{1}{0.500} = \mathbf{2.00}$

$\tan \theta = \dfrac{\sin \theta}{\cos \theta} = \dfrac{0.625}{0.500} = \mathbf{1.25}$

$\cot \theta = \dfrac{\cos \theta}{\sin \theta} = \dfrac{0.500}{0.625} = \mathbf{0.80}$

*Problem 6* If cosec $\theta = 1.4622$ determine the values of sec $\theta$ and cot $\theta$. (Assume $\theta$ is an acute angle.)

If cosec $\theta = 1.4622$ then $\theta = \operatorname{arccosec} 1.4622 = 43° \, 9'$.
From tables or calculator, **sec $43° \, 9' = 1.3707$** and cot $43° \, 9' = \mathbf{1.0668}$

*Problem 7* Evaluate the following expression, correct to 4 significant figures:
$\dfrac{4 \sec 32° \, 10' - 2 \cot 15° \, 19'}{3 \operatorname{cosec} 63° \, 8' \tan 14° \, 57'}$

From tables or calculator: $\sec 32° \, 10' = 1.1814$, $\cot 15° \, 19' = 3.6513$,
$\operatorname{cosec} 63° \, 8' = 1.1210$, $\tan 14° \, 57' = 0.2670$

Hence $\dfrac{4 \sec 32° \, 10' - 2 \cot 15° \, 19'}{3 \operatorname{cosec} 63° \, 8' \tan 14° \, 57'} = \dfrac{4(1.1814) - 2(3.6513)}{3(1.1210)(0.2670)}$

$= \dfrac{4.7256 - 7.3026}{0.8979} = \dfrac{-2.5770}{0.8979} = \mathbf{-2.870}$, correct to 4 significant figures

*Problem 8* A surveyor measures the angle of elevation of the top of a building as $17° \, 15'$. He moves 80.0 m nearer to the building and measures the angle of elevation at $29°\, 35'$. Determine the perpendicular height of the building.

In *Fig 9*, AB represents the perpendicular height of the building, C is the initial position and D the final position of the surveyor.

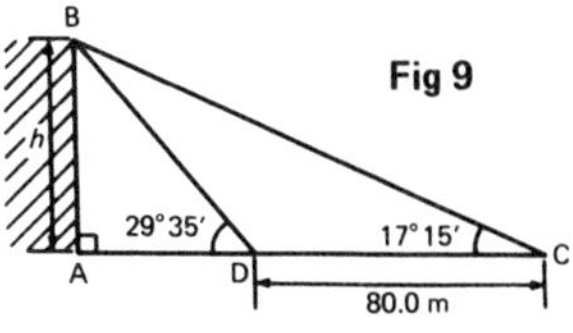

Using trigonometric ratios, $\cot 17° \, 15' = \dfrac{AC}{h}$,
from which $AC = h \cot 17° \, 15'$.
Similarly, $AD = h \cot 29° \, 35'$.
Since $DC = AC - AD$,
$DC = h \cot 17° \, 15' - h \cot 29° \, 35'$
$= h(\cot 17° \, 15' - \cot 29° \, 35') = h(1.4594)$
Hence the perpendicular height of the building, $h = \dfrac{80.0}{1.4594} = \mathbf{54.82 \text{ m}}$

**Fig 9**

*Problem 9* Evaluate (i) sin 150°, (ii) tan 150° and (iii) sec 150°, using 4 figure tables

Following the procedure of para 5(vii):
(a) A sketch is shown in *Fig 10..*
(b) An angle of 150° lies in the second quadrant.
(c) The acute angle $\theta$ measured to the horizontal is $180° - 150° = 30°$.
(d) From tables, sin 30° = 0.5000, tan 30° = 0.5774 and sec 30° = 1.1547
(e) Only sine and cosecant are positive in the second quadrant hence:
   (i)   sin 150° = +sin 30° = **+0.5000**
   (ii)  tan 150° = −tan 30° = **−0.5774**
   (iii) sec 150° = −sec 30° = **−1.1547**

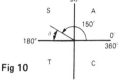

**Fig 10**

*Problem 10* Evaluate, using 4-figure tables, (a) cosine 251° 19′; (b) cosec 251° 19′; (c) cot 251° 19′.

From *Fig 11*, $\theta = 251° 19' - 180° = 71° 19'$ and from tables:
cos 71° 19′ = 0.3203, cosec 71° 19′ = 1.0556
and cot 71° 19′ = 0.3382
In the third quadrant only tangent and cotangent are positive.
Hence (a) cos   251° 19′ = −cos   71° 19′ = **−0.3203**
      (b) cosec 251° 19′ = −cosec 71° 19′ = **−1.0556**
      (c) cot   251° 19′ = +cot   71° 19′ = **+0.3382**

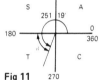

**Fig 11**

*Problem 11* Evaluate, using a calculator, (a) tan 146° 32′; (b) sec 423° 11′; correct to 4 decimal places.

(a) tan 146° 32′ = tan $146\frac{32}{60}$ .  $\frac{32}{60}$ = 0.5333333

   Hence tan 146.5333° = **−0.6610**, correct to 4 decimal places.

(b) sec 423° 11′ = sec $423\frac{11}{60}$ = sec 423.18333°

   cos 423.18333° = 0.4511372

   Hence sec 423.18333° = $\frac{1}{0.4511372}$ = **2.2166**, correct to 4 decimal places.

(*Note*: sec 423° 11′ is exactly the same as sec (423° 11′−360°), i.e. sec 63° 11′.)

*Problem 12* Evaluate (a) sin (−123° 17′); (b) cot (−123° 17′).

A negative angle means that the angle is measured in a clockwise direction. $-123°\ 17'$ is shown in *Fig 12* and lies in the third quadrant.
$\theta = 180° - 123°\ 17' = 56°\ 43'$.
$\sin 56°\ 43' = 0.8360$ and $\cot 56°\ 43' = 0.6565$.
In the third quadrant only tangent and cotangent are positive.
Hence
(a) $\sin(-123°\ 17') = -\sin 56°\ 43' = \mathbf{-0.8360}$
(b) $\cot(-123°\ 17') = +\cot 56°\ 43' = \mathbf{+0.6565}$

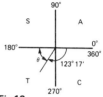

**Fig 12**

*Problem 13* Evaluate (a) $\cos\dfrac{5\pi}{24}$; (b) $\operatorname{cosec}\dfrac{3\pi}{4}$

(a) Angles stated in terms of $\pi$ are measured in radians.
$360° = 2\pi$ radians, from which $180° = \pi$ radians.

Hence $\dfrac{5\pi}{24}$ rad $= \dfrac{5}{24}(180)° = 37.5°$

$\cos 37.5° = \cos 37°\ 30' = \mathbf{0.7934}$

(b) $\dfrac{3\pi}{4}$ rad $= \dfrac{3}{4}(180)° = 135°$.

$\operatorname{cosec} 135° = +\operatorname{cosec} 45°$ (since sine and cosecant are positive in the second quadrant)
$= \mathbf{+1.4142}$

*Problem 14* Use tables to show that $\operatorname{cosec} 204°\ 10' = \dfrac{1}{\sin 204°\ 10'}$, correct to 3 decimal places.

$\operatorname{Cosec} 204°\ 10' = -\operatorname{cosec} 24°\ 10'$ (since cosecant is negative in the third quadrant)
$= -2.4428 = -2.443$ correct to 3 decimal places.
$\sin 204°\ 10' = -\sin 24°\ 10' = -0.4094$ (since sine is negative in the third quadrant)
From reciprocal tables $\dfrac{1}{-0.4094} = -\dfrac{1}{4.094 \times 10^{-1}} = -\left(\dfrac{1}{4.094}\right)(10)$
$= -(0.2443)(10)$
$= -2.443$

Hence $\operatorname{cosec} 204°\ 10' = \dfrac{1}{\sin 204°\ 10'}$.

*Problem 15* Determine all the angles between 0° and 360° (a) whose sine is $-0.4638$ and (b) whose tangent is $1.7629$.

(a) The angles whose sine is −0.4638 occurs in the third and fourth quadrants since sine is negative in these quadrants (see *Fig 13(a)*).
From *Fig 13(b)*, θ = arcsin 0.4638 = 27° 38'.
Measured from 0°, the two angles between 0° and 360° whose sine is −0.4638 are 180° + 27° 38', i.e. **207° 38'** and 360° − 27° 38', i.e. **332° 22'**.
(Note that a calculator generally only gives one answer, i.e. −27.632588°.)

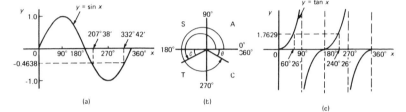

**Fig 13**

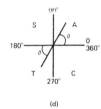

(d)

(b) A tangent is positive in the first and third quadrants (see *Fig 13(c)*).
From *Fig 13(d)*, θ = arctan 1.7629 = 60° 26'.
Measured from 0°, the two angles between 0° and 360° whose tangent is 1.7629 are **60° 26'** and 180° + 60° 26', i.e. **240° 26'**.

*Problem 16* Solve for angles of α between 0° and 360°: (a) arcsec −2.1499 = α; (b) arccot 1.3111 = α.

(a) Secant is negative in the second and third quadrants (i.e. the same as for cosine). From *Fig 14(a)*, θ = arcsec 2.1499 = 62° 17'. Measured from 0°, the two angles between 0° and 360° whose secant is −2.1499 are
α = 180° − 62° 17' = **117° 43'**
and α = 180° + 62° 17' = **242° 17'**
(b) Cotangent is positive in the first and third quadrants (i.e. same as for tangent). From *Fig 14(b)*, θ = arccot 1.3111 = 37° 20'.
Hence α = **37° 20'** and 180° + 37° 20' = **217° 20'**.

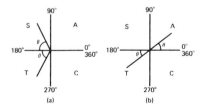

**Fig 14**

## C. FURTHER PROBLEMS ON TRIGONOMETRIC RATIOS FOR ANGLES OF ANY MAGNITUDE

1 For the right-angled triangle shown in *Fig 15*, find (a) sin $\alpha$; (b) cos $\theta$; (c) sec $\theta$; (d) cosec $\alpha$; (e) tan $\theta$; (f) cot $\theta$; (g) cot $\alpha$

Fig 15

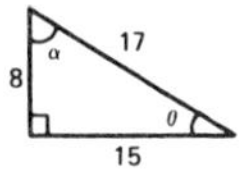

$\left[\text{(a) } \frac{15}{17}; \text{ (b) } \frac{15}{17}; \text{ (c) } 1\frac{2}{15}; \text{ (d) } 1\frac{2}{15}; \text{ (e) } \frac{8}{15}; \text{ (f) } 1\frac{7}{8}; \text{ (g) } \frac{8}{15}\right]$

2 If $\tan\theta = \frac{7}{24}$, find the other five trigonometric ratios in fraction form.

$\left[\sin\theta = \frac{7}{25}; \cos\theta = \frac{24}{25}; \sec\theta = 1\frac{1}{24}; \text{cosec}\,\theta = 3\frac{4}{7}; \cot\theta = 3\frac{3}{7}\right]$

3 In a right-angled triangle PQR, PQ = 11.6 cm, QR = 5.3 cm and $R = 90°$. Evaluate, correct to 4 significant figures, (a) sec $Q$; (b) cot $P$; (c) tan $Q$; (d) cosec $Q$; (e) cot $Q$.

[(a) 2.189; (b) 1.947; (c) 1.947; (d) 1.124; (e) 0.5137]

In *Problems 4 to 6*, use 4-figure tables to evaluate.

4 (a) sec 15° 8′; (b) sec 47° 29′; (c) sec 81° 18′

[(a) 1.0360; (b) 1.4797; (c) 6.6111]

5 (a) cosec 25° 32′; (b) cosec 58° 3′; (c) cosec 79° 52′

[(a) 2.3200; (b) 1.1786; (c) 1.0159]

6 (a) cot 15° 23′; (b) cot 45° 9′; (c) cot 81° 47′;

[(a) 3.6350; (b) 1.9948; (c) 0.1444]

In *Problems 7 to 9*, use 4-figure tables to evaluate the acute angles.

7 (a) arcsec 1.0091; (b) arcsec 1.3527; (c) arcsec 2.6317.

[(a) 7° 42′; (b) 42° 20′; (c) 67° 40′]

8 (a) arccosec 2.3781; (b) arccosec 1.6217; (c) arccosec 1.0710.

[(a) 24° 52′; (b) 38° 4′; (c) 69° 1′]

9 (a) arccot 3.4236; (b) arccot 2.3148; (c) arccot 0.4615.

[(a) 16° 17′; (b) 23° 22′; (c) 65° 14′]

10 If $\sin\theta = 0.60$ and $\cos\theta = 0.80$, find sec $\theta$, cosec $\theta$, tan $\theta$ and cot $\theta$, without using trigonometrical tables, correct to 3 decimal places.

[1.250; 1.667; 0.750; 1.333]

11 If $\tan x = 1.5276$ determine sec $x$, cosec $x$ and cot $x$. (Assume $x$ is an acute angle.)

[1.8258; 1.1952; 0.6546]

12 Evaluate, using 4 figure tables, correct to 4 significant figures.

(a) 3 cot 14° 15′ sec 23° 9′

(b) $\dfrac{\text{cosec } 27°\ 19' + \sec 45°\ 29'}{1 - \text{cosec } 27°\ 19' \sec 45°\ 29'}$

(c) $\dfrac{30 \tan 61° \sec 54° - 15 \cot 14°}{2\ \text{cosec } 24°}$

[(a) 12.85; (b) −1.710; (c) 6.490]

13 From the top of a vertical cliff 80.0 m high the angles of depression of two buoys lying due west of the cliff are 23° and 15° respectively. How far are the buoys apart?

[110.1 m]

14 Evaluate (a) sin 148°; (b) sin 236°; (c) sin 342° 13′.
[(a) 0.5299; (b) −0.8290; (c) −0.3054]

15 Evaluate (a) cos 82°; (b) cos 171° 29′; (c) cos 302°.
[(a) 0.1392; (b) −0.9890; (c) 0.5299]

16 Evaluate (a) tan 111°; (b) tan 211°; (c) tan 311° 11′.
[(a) −2.6051; (b) 0.6009; (c) −1.1430]

17 Evaluate (a) sec 125°; (b) sec 204°; (c) sec 297° 8′.
[(a) −1.7434; (b) −1.0946; (c) 2.1927]

18 Evaluate (a) cosec 103°; (b) cosec 211° 8′; (c) cosec 347°.
[(a) 1.0263; (b) −1.9341; (c) −4.4454]

19 Evaluate (a) cot 138°; (b) cot 220° 50′; (c) cot 316°.
[(a) −1.1106; (b) 1.1571; (c) −1.0355]

20 Evaluate (a) cos 431° 19′; (b) cot 511°; (c) sec 461° 29′.
[(a) 0.3203; (b) −1.8040; (c) −5.0230]

21 Evaluate (a) cosec (−125°); (b) tan (−241°); (c) sec (−49° 15′).
[(a) −1.2208; (b) −1.8040; (c) 1.5320]

22 Evaluate (a) sin $3\pi/8$; (b) sec $5\pi/16$; (c) cosec $4\pi/9$.
[(a) 0.9239; (b) 1.8000; (c) 1.0154]

23 Use tables to show that sec 146° 10′ $\dfrac{1}{\cos 146° 10'}$

24 Find all the angles between 0° and 360°: (a) whose sine is −0.7321; (b) whose cosecant is 2.5317; (c) whose cotangent is −0.6312.
[(a) 227° 4′ or 312° 56′; (b) 23° 16′ or 156° 44′; (c) 122° 16′ or 302° 16′]

25 Solve for all values of $\theta$ between 0° and 360°: (a) arccos −0.5316 = $\theta$;
(b) arcsec 2.3162 = $\theta$; (c) arctan 0.8314 = $\theta$.
[(a) 122° 7′ or 237° 53′; (b) 64° 25′ or 295° 35′; (c) 39° 44′ or 219° 44′]

26 Use a calculator to check the answers to *Problems 3 to 25*.

# 2 The solution of triangles and their areas

## A. MAIN POINTS CONCERNED WITH THE SOLUTION OF TRIANGLES AND THEIR AREAS

1  To '**solve a triangle**' means 'to find the values of unknown sides and angles'.
   If a triangle is **right-angled**, trigonometric ratios and the theorem of Pythagoras may be used for its solution. However, for a **non-right-angled triangle**, trigonometric ratios and Pythagoras' theorem **cannot** be used. Instead, two rules, called the sine rule and the cosine rule, are used.

2  With reference to triangle ABC of *Fig 1*, the **sine rule** states:

$$\frac{a}{\sin A} = \frac{b}{\sin B} = \frac{c}{\sin C}$$

   The rule may be used only when:
   (i) 1 side and any 2 angles are initially given, or
   (ii) 2 sides and an angle (not the included angle) are initially given.

3  With reference to triangle ABC of *Fig 1*, the **cosine rule** states:
   $a^2 = b^2 + c^2 - 2bc \cos A$
   or $b^2 = a^2 + c^2 - 2ac \cos B$
   or $c^2 = a^2 + b^2 - 2ab \cos C$

   The rule may be used only when:
   (i) 2 sides and the included angle are initially given, or
   (ii) 3 sides are initially given.

4  The **area of any triangle** such as ABC of *Fig 1* is given by:

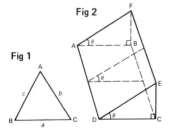

   (i) $\frac{1}{2} \times$ base $\times$ perpendicular height or

   (ii) $\frac{1}{2} ab \sin C$ or
   $\frac{1}{2} ac \sin B$ or $\frac{1}{2} bc \sin A$

   or (iii) $\sqrt{[s(s-a)(s-b)(s-c)]}$,

   where $s = \frac{a+b+c}{2}$

5  **Lengths and areas on an inclined plane**
   In *Fig 2*, rectangle ADEF is a plane inclined at an angle of $\theta$ to the horizontal plane ABCD.

   $\cos \theta = \frac{DC}{DE}$, from which, $DE = \frac{DC}{\cos \theta}$

13

Hence the line of greatest slope on an inclined plane is given by:

$\left(\dfrac{1}{\cos \theta}\right)$ (its projection on to the horizontal plane).

Area of ADEF = (AD)(DE) = (AD) $\left(\dfrac{DC}{\cos \theta}\right)$ = $\left(\dfrac{1}{\cos \theta}\right)$ (area of horizontal plane)

(see problems 7 and 8)

6 There are a number of **practical situations** where the use of trigonometry is needed to find unknown sides and angles of triangles. (See *Problems 9 to 15*.)

## B. WORKED PROBLEMS ON THE SOLUTION OF TRIANGLES AND THEIR AREAS

*Problem 1* In the triangle XYZ, $X = 51°$, $Y = 67°$ and YZ = 15.2 cm. Solve the triangle and find its area.

The triangle XYZ is shown in *Fig 3*.
Since the angles in a triangle add up to 180°, then
$Z = 180° - 51° - 67° = 62°$.

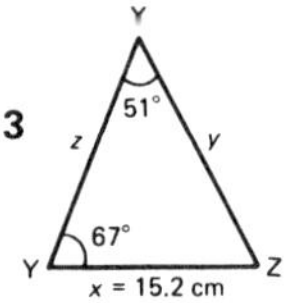

**Fig 3**

Applying the sine rule: $\dfrac{15.2}{\sin 51°} = \dfrac{y}{\sin 67°} = \dfrac{z}{\sin 62°}$

Using $\dfrac{15.2}{\sin 51°} = \dfrac{y}{\sin 67°}$ and transposing gives: $y = \dfrac{15.2 \sin 67°}{\sin 51°} = 18.00$ cm = XZ.

Using $\dfrac{15.2}{\sin 51°} = \dfrac{z}{\sin 62°}$ and transposing gives: $z = \dfrac{15.2 \sin 62°}{\sin 51°} = 17.27$ cm = XY.

Area of triangle XYZ = $\dfrac{1}{2}xy \sin Z = \dfrac{1}{2}(15.2)(18.00) \sin 62° = 120.8$ cm$^2$

(or area = $\dfrac{1}{2}xz \sin Y = \dfrac{1}{2}(15.2)(17.27) \sin 67° = 120.8$ cm$^2$).

It is always worth checking with triangle problems that the longest side is opposite the largest angle, and vice-versa. In this problem, $Y$ is the largest angle and thus XZ should be the longest of the three sides. (In this problem, and in the following problems, the method of calculation is not specified—logarithms, including logarithms of sines, or calculators may be used.)

*Problem 2* Solve the triangle ABC given $B = 78° 51'$, AC = 22.31 mm and AB = 17.92 mm. Find also its area.

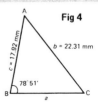

**Fig 4**

Triangle ABC is shown in *Fig 4*.
Applying the sine rule: $\dfrac{22.31}{\sin 78° 51'} = \dfrac{17.92}{\sin C}$,

from which, $\sin C = \dfrac{17.92 \sin 78° 51'}{22.31} = 0.7881$

Hence $C = \arcsin 0.7881 = 52°\ 0'$ or $128°\ 0'$ (see chapter 1).
Since $B = 78°\ 51'$, $C$ cannot be $128°\ 0'$, since $128°\ 0' + 78°\ 51'$ is greater than $180°$.
Thus only $C = 52°\ 0'$ is valid.
Angle $A = 180° - 78°\ 51' - 52°\ 0' = 49°\ 9'$.

Applying the sine rule: $\dfrac{a}{\sin 49°\ 9'} = \dfrac{22.31}{\sin 78°\ 51'}$,

from which, $a = \dfrac{22.31 \sin 49°\ 9'}{\sin 78°\ 51'} = 17.20$ mm

Hence $A = 49°\ 9'$, $C = 52°\ 0'$ and **BC = 17.20 mm**

Area of triangle ABC = $\dfrac{1}{2} ac \sin B = \dfrac{1}{2}(17.20)(17.92) \sin 78°\ 51' =$ **151.2 mm²**

---

*Problem 3* Solve the triangle PQR and find its area given that QR = 36.5 mm, PR = 29.6 mm and $Q = 36°$.

---

Triangle PQR is shown in *Fig 5*.

Applying the sine rule: $\dfrac{29.6}{\sin 36°} = \dfrac{36.5}{\sin P}$

from which, $\sin P = \dfrac{36.5 \sin 36°}{29.6} = 0.7248$

Hence $P = \arcsin 0.7248 = 46°\ 27'$ or $133°\ 33'$.
When $P = 46°\ 27'$ and $Q = 36°$ then $R = 180° - 46°\ 27' - 36° = 97°\ 33'$.
When $P = 133°\ 33'$ and $Q = 36°$ then $R = 180° - 133°\ 33' - 36° = 10°\ 27'$.
Thus, in this problem, there are **two** separate sets of results and both are feasible solutions. Such a situation is called the **ambiguous case**.

*Case 1* $P = 46°\ 27'$, $Q = 36°$, $R = 97°\ 33'$, $p = 36.5$ mm and $q = 29.6$ mm.

From the sine rule: $\dfrac{r}{\sin 97°\ 33'} = \dfrac{29.6}{\sin 36°}$

**Fig 5**

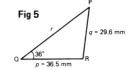

from which, $r = \dfrac{29.6 \sin 97°\ 33'}{\sin 36°} =$ **49.92 mm**.

Area = $\dfrac{1}{2} pq \sin R = \dfrac{1}{2}(36.5)(29.6) \sin 97°\ 33' =$ **535.5 mm²**.

*Case 2* $P = 133°\ 33'$, $Q = 36°$, $R = 10°\ 27'$, $p = 36.5$ mm and $q = 29.6$ mm.

From the sine rule: $\dfrac{r}{\sin 10°\ 27'} = \dfrac{29.6}{\sin 36°}$

**Fig 6**

from which, $r = \dfrac{29.6 \sin 10°\ 27'}{\sin 36°} =$ **9.134 mm**.

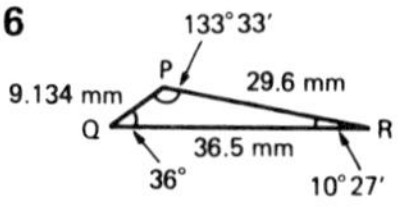

Area = $\dfrac{1}{2} pq \sin R = \dfrac{1}{2}(36.5)(29.6) \sin 10°\ 27' =$ **97.98 mm²**.

Triangle PQR for case 2 is shown in *Fig 6*.

*Problem 4* Solve triangle DEF and find its area given that EF = 35.0 mm, DE = 25.0 mm and $E = 64°$.

Triangle DEF is shown in *Fig 7*. Applying the cosine rule:

$e^2 = d^2 + f^2 - 2df \cos E$

i.e. $e^2 = (35.0)^2 + (25.0)^2 - \{2(35.0)(25.0) \cos 64°\}$

$= 1225 + 625 - 767.1 = 1083$

$e = \sqrt{1083} = $ **32.91 mm.**

Applying the sine rule: $\dfrac{32.91}{\sin 64°} = \dfrac{25.0}{\sin F}$,

from which, $\sin F = \dfrac{25.0 \sin 64°}{32.91} = 0.6828$

Thus $F = \arcsin 0.6828 = 43° \ 4'$ or $136° \ 56'$.

$F = 136° \ 56'$ is not possible in this case since $136° \ 56' + 64°$ is greater than $180°$. Thus only $F = $ **43° 4'** is valid.

$D = 180° - 64° - 43° \ 4' = $ **72° 56'**.

Area of triangle DEF $= \dfrac{1}{2} df \sin E = \dfrac{1}{2}(35.0)(25.0) \sin 64° = $ **393.2 mm²**.

---

*Problem 5* A triangle ABC has sides $a = 9.0$ cm, $b = 7.5$ cm and $c = 6.5$ cm. Determine its three angles and its area.

Triangle ABC is shown in *Fig 8*. It is usual to firstly calculate the largest angle to determine whether the triangle is acute or obtuse. In this case the largest angle is A (i.e. opposite the longest side).

Applying the cosine rule: $a^2 = b^2 + c^2 - 2bc \cos A$

from which, $2bc \cos A = b^2 + c^2 - a^2$

and $\cos A = \dfrac{b^2 + c^2 - a^2}{2bc} = \dfrac{7.5^2 + 6.5^2 - 9.0^2}{2(7.5)(6.5)}$

$= 0.1795$

Hence $A = \arccos 0.1795 = $ **79° 40'** (or $280° \ 20'$, which is obviously impossible). The triangle is thus acute angled since $\cos A$ is positive. (If $\cos A$ had been negative, angle $A$ would be obtuse, i.e. lie between $90°$ and $180°$.)

Applying the sine rule: $\dfrac{9.0}{\sin 79° 40'} = \dfrac{7.5}{\sin B}$

from which, $\sin B = \dfrac{7.5 \sin 79° 40'}{9.0} = 0.8198$

Hence $B = \arcsin 0.8198 = $ **55° 4'**.

$C = 180° - 79° \ 40' - 55° \ 4' = $ **45° 16'**.

Area $= \sqrt{[s(s-a)(s-b)(s-c)]}$, where $s = \dfrac{a+b+c}{2} = \dfrac{9.0+7.5+6.5}{2} = 11.5$ cm.

Hence area $= \sqrt{[11.5(11.5-9.0)(11.5-7.5)(11.5-6.5)]} = \sqrt{[11.5(2.5)(4.0)(5.0)]}$

$= $ **23.98 cm²**.

Alternatively, area $= \dfrac{1}{2} ab \sin C = \dfrac{1}{2}(9.0)(7.5) \sin 45° 16' = $ **23.98 cm²**

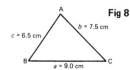
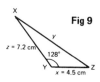

*Problem 6* Solve triangle XYZ (*Fig 9*) and find its area given that $Y = 128°$, XY = 7.2 cm and YZ = 4.5 cm.

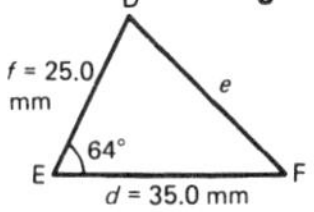

Applying the cosine rule:
$$\begin{aligned} y^2 &= x^2+z^2-2xz\cos Y \\ &= (4.5)^2+(7.2)^2-\{2(4.5)(7.2)\cos 128°\} \\ &= 20.25+51.84-\{-39.89\} \\ &\quad \text{(since } \cos 128° \equiv -\cos 52°) \\ &= 20.25+51.84+39.89 = 112.0 \\ y &= \sqrt{(112.0)} = \textbf{10.58 cm.} \end{aligned}$$

Applying the sine rule: $\dfrac{10.58}{\sin 128°} = \dfrac{7.2}{\sin Z}$

from which, $\sin Z = \dfrac{7.2 \sin 128°}{10.58} = 0.5363$.

Hence $Z = \arcsin 0.5363 = \textbf{32° 26}'$ (or $147° 34'$ which, here, is impossible).
$X = 180° - 128° - 32° 26' = \textbf{19° 34}'$.

Area $= \dfrac{1}{2}xz \sin Y = \dfrac{1}{2}(4.5)(7.2)\sin 128° = \textbf{12.77 cm}^2$

*Problem 7* A vertical, cylindrical ventilation shaft of diameter 36.0 cm has its end at an angle of 20° to the horizontal as shown in *Fig 10*. Determine the area of the end cover plate.

Cover plate area $= \left(\dfrac{1}{\cos \theta}\right)$ (horizontal plane area), from para. 5,

$= \left(\dfrac{1}{\cos 20°}\right)\left[\pi\left(\dfrac{36.0}{2}\right)^2\right] = \textbf{1083 cm}^2$

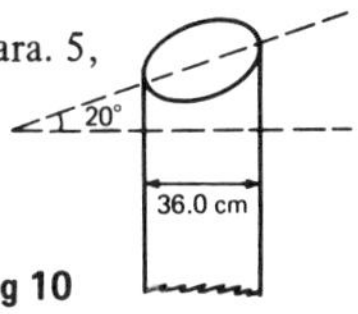

**Fig 10**

*Problem 8* A rectangular chimney stack having dimensions of 1.2 m by 0.70 m passes through a roof that has a pitch of 35°. Determine the area of the void in the roof through which the stack passes.

Area of void in roof $= \left(\dfrac{1}{\cos \theta}\right)$ (area of horizontal plane)

$= \left(\dfrac{1}{\cos 35°}\right)(1.2 \times 0.70) = \textbf{1.025 m}^2$.

*Problem 9* A room 8.0 m wide has a span roof which slopes at 33° on one side and 40° on the other. Find the length of the roof slopes, correct to the nearest centimetre.

A section of the roof is shown in *Fig 11*.
Angle at ridge, $B = 180° - 33° - 40° = 107°$.

From the sine rule: $\dfrac{8.0}{\sin 107°} = \dfrac{a}{\sin 33°}$

from which, $a = \dfrac{8.0 \sin 33°}{\sin 107°} = 4.556$ m.

Also from the sine rule: $\dfrac{8.0}{\sin 107°} = \dfrac{c}{\sin 40°}$

from which, $c = \dfrac{8.0 \sin 40°}{\sin 107°} = 5.377$ m.

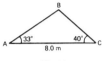

**Fig 11**

Hence the roof slopes are 4.56 m and 5.38 m, correct to the nearest centimetre.

*Problem 10* A man leaves a point walking at 6.5 km/h in a direction E 20° N (i.e. a bearing of 70°). A cyclist leaves the same point at the same time in a direction E 40° S (i.e. a bearing of 130°) travelling at a constant speed. Find the average speed of the cyclist if the walker and cyclist are 80 km apart after 5 hours.

After 5 hours the walker has travelled $5 \times 6.5 = 32.5$ km (shown as AB in *Fig 12*). If AC is the distance the cyclist travels in 5 hours then BC = 80 km.

Applying the sine rule: $\dfrac{80}{\sin 60°} = \dfrac{32.5}{\sin C}$

from which, $\sin C = \dfrac{32.5 \sin 60°}{80} = 0.3518$

Hence $C = \arcsin 0.3518 = 20°\ 36'$ (or $159°\ 24'$ which is impossible in this case).
$B = 180° - 60° - 20°\ 36' = 99°\ 24'$

Applying the sine rule: $\dfrac{80}{\sin 60°} = \dfrac{b}{\sin 99°\ 24'}$

from which, $b = \dfrac{80 \sin 99°\ 24'}{\sin 60°} = 91.14$ km

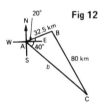

**Fig 12**

Since the cyclist travels 91.14 km in 5 hours then:

average speed = $\dfrac{\text{distance}}{\text{time}} = \dfrac{91.14}{5} = 18.23$ km/h

*Problem 11* Two voltage phasors are shown in *Fig 13*. If $V_1 = 40$ V and $V_2 = 100$ V determine the value of their resultant (i.e. length OA) and the angle the resultant makes with $V_1$.

Angle OBA = 180° − 45° = 135°.
Applying the cosine rule: 
$$OA^2 = V_1^2 + V_2^2 - 2V_1 V_2 \cos OBA$$
$$= 40^2 + 100^2 - \{2(40)(100) \cos 135°\}$$
$$= 1600 + 10\,000 - \{-5657\}$$
$$= 1600 + 10\,000 + 5657 = 17\,257$$
The resultant OA $= \sqrt{(17\,257)} = 131.4$ V

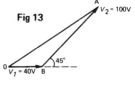

**Fig 13**

Applying the sine rule: $\dfrac{131.4}{\sin 135°} = \dfrac{100}{\sin AOB}$

from which, $\sin AOB = \dfrac{100 \sin 135°}{131.4} = 0.5381$

Hence angle AOB = arcsin 0.5381 = 32° 33′ (or 147° 27′ which is impossible in this case).

**Hence the resultant voltage is 131.4 volts at 32° 33′ to $V_1$**

---

*Problem 12* In *Fig 14*, PR represents the inclined jib of a crane and is 10.0 m long. PQ is 4.0 m long. Determine the length of tie QR and the inclination of the jib to the vertical.

---

Applying the sine rule: $\dfrac{PR}{\sin 120°} = \dfrac{PQ}{\sin R}$

from which, $\sin R = \dfrac{PQ \sin 120°}{PR} = \dfrac{(4.0) \sin 120°}{10.0} = 0.3464$

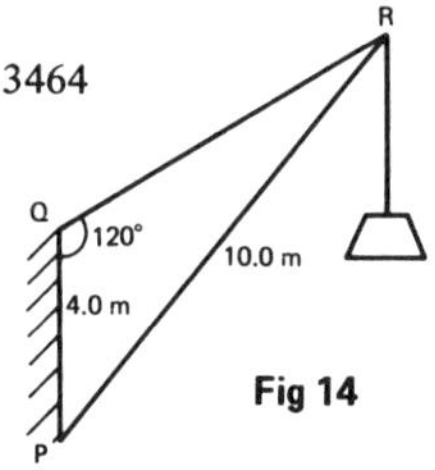

Hence $R$ = arcsin 0.3464 = 20° 16′
(or 159° 44′, which is impossible in this case).
$P = 180° - 120° - 20° 16′ = 39° 44′$,
**which is the inclination of the jib to the vertical.**

Applying the sine rule: $\dfrac{10.0}{\sin 120°} = \dfrac{QR}{\sin 39° 44′}$

**Fig 14**

from which, $QR = \dfrac{10.0 \sin 39° 44′}{\sin 120°} = 7.38$ m = **length of tie**

---

*Problem 13* A vertical aerial stands on horizontal ground. A surveyor positioned due east of the aerial measures the elevation of the top as 48°. He moves due south 30.0 m and measures the elevation as 44°. Determine the height of the aerial.

---

In *Fig 15*, DC represents the aerial, A is the initial position of the surveyor and B his final position.
From triangle ACD, tan 48° = DC/AC, from which, AC = DC/tan 48° = DC cot 48°.
Similarly, from triangle BCD, BC = DC cot 44°.
For triangle ABC, using Pythagoras' theorem: $BC^2 = AB^2 + AC^2$

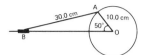

Fig 16

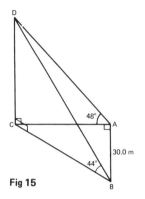

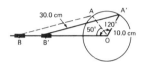

Fig 15

Fig 17

$(DC \cot 44°)^2 = (30.0)^2 + (DC \cot 48°)^2$

$DC^2(\cot^2 44° - \cot^2 48°) = 30.0^2$

$DC^2 = \dfrac{30.0^2}{\cot^2 44° - \cot^2 48°} = 3440$

Hence, height of aerial, $DC = \sqrt{3440} = 58.65$ m

*Problem 14* A crank mechanism of a petrol engine is shown in *Fig 16*. Arm OA is 10.0 cm long and rotates clockwise about 0. The connecting rod AB is 30.0 cm long and end B is constrained to move horizontally.

(a) For the position shown in *Fig 16* determine the angle between the connecting rod AB and the horizontal and the length of OB.
(b) How far does B move when angle AOB changes from 50° to 120°?

(a) Applying the sine rule: $\dfrac{AB}{\sin 50°} = \dfrac{AO}{\sin B}$

from which, $\sin B = \dfrac{AO \sin 50°}{AB} = \dfrac{10.0 \sin 50°}{30.0} = 0.2553$

Hence $B = \arcsin 0.2553 = 14° \ 47'$ (or 165° 13′, which is impossible in this case).

**Hence the connecting rod AB makes an angle of 14° 47′ with the horizontal.**
Angle $OAB = 180° - 50° - 14° \ 47' = 115° \ 13'$.

Applying the sine rule: $\dfrac{30.0}{\sin 50°} = \dfrac{OB}{\sin 115° \ 13'}$

from which, $OB = \dfrac{30.0 \sin 115° \ 13'}{\sin 50°} = $ **35.43 cm**

(b) *Fig 17* shows the initial and final positions of the crank mechanism.

In triangle OA′B′, applying the sine rule: $\dfrac{30.0}{\sin 120°} = \dfrac{10.0}{\sin A'B'O}$

from which, sin A'B'O = $\frac{10.0 \sin 120°}{30.0}$ = 0.2887

Hence A'B'O = arcsin 0.2887 = 16° 47' (or 163° 13' which is impossible in this case).

Angle OA'B' = 180° − 120° − 16° 47' = 43° 13'.

Applying the sine rule: $\frac{30.0}{\sin 120°} = \frac{OB'}{\sin 43° 13'}$ ,

from which, OB' = $\frac{30.0 \sin 43° 13'}{\sin 120°}$ = 23.72 cm.

Since OB = 35.43 cm and OB' = 23.72 cm then BB' = 35.43 − 23.72 = 11.68 cm.
**Hence B moves 11.68 cm when angle AOB changes from 50° to 120°.**

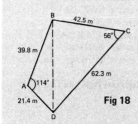

*Problem 15*  The area of a field is in the form of a quadrilateral ABCD as shown in *Fig 18*. Determine its area.

**Fig 18**

A diagonal drawn from B to D divides
the quadrilateral into two triangles.
Area of quadrilateral ABCD
= area of triangle ABD + area of triangle BCD
= $\frac{1}{2}$(39.8)(21.4) sin 114° + $\frac{1}{2}$(42.5)(62.3) sin 56°
= 389.04 + 1097.5
= **1487 m²**

## C. FURTHER PROBLEMS ON THE SOLUTION OF TRIANGLES AND THEIR AREAS

1  Use the sine rule to solve the following triangles ABC and find their areas.
 (a) $A = 29°$; $B = 68°$; $b = 27$ mm
 (b) $B = 71° 26'$; $C = 56° 32'$; $b = 8.60$ cm.
 (c) $A = 117°$; $C = 24° 30'$; $a = 15.2$ mm.

$$\begin{bmatrix} \text{(a) } C = 83°; \ a = 14.1 \text{ mm}; \ c = 28.9 \text{ mm}; \text{ area} = 189 \text{ mm}^2 \\ \text{(b) } A = 52° 2'; \ c = 7.568 \text{ cm}; \ a = 7.152 \text{ cm}; \text{ area} = 25.65 \text{ cm}^2 \\ \text{(c) } B = 38° 30'; \ b = 10.62 \text{ mm}; \ c = 7.074 \text{ mm}; \text{ area} = 33.47 \text{ mm}^2 \end{bmatrix}$$

2. Use the sine rule to solve the following triangles DEF and find their areas.
   (a) $d = 17$ cm; $f = 22$ cm; $F = 26°$
   (b) $e = 4.20$ m; $f = 7.10$ m; $F = 81°$
   (c) $d = 32.6$ mm; $e = 25.4$ mm; $D = 104° 22'$

   $$\begin{bmatrix} \text{(a) } D = 19° 48'; E = 134° 12'; e = 36.0 \text{ cm; area} = 134 \text{ cm}^2 \\ \text{(b) } E = 35° 45'; D = 63° 15'; d = 6.419 \text{ m; area} = 13.31 \text{ m}^2 \\ \text{(c) } E = 49° 0'; F = 26° 38'; f = 15.09 \text{ mm; area} = 185.6 \text{ mm}^2 \end{bmatrix}$$

3. Use the sine rule to solve the following triangles JKL and find their areas.
   (a) $j = 3.85$ cm; $k = 3.23$ cm; $K = 36°$
   (b) $k = 46$ mm; $l = 36$ mm; $L = 35°$
   (c) $j = 2.92$ m; $l = 3.24$ m; $J = 27° 30'$

   $$\begin{bmatrix} \text{(a) } J = 44° 29'; L = 99° 31'; l = 5.420 \text{ cm; area} = 6.133 \text{ cm}^2 \\ \text{OR } J = 135° 31'; L = 8° 29'; l = 0.811 \text{ cm; area} = 0.917 \text{ cm}^2 \\ \text{(b) } K = 47° 8'; J = 97° 52'; j = 62.2 \text{ mm; area} = 820.6 \text{ mm}^2 \\ \text{OR } K = 132° 52'; J = 12° 8'; j = 13.19 \text{ mm; area} = 174.0 \text{ mm}^2 \\ \text{(c) } L = 30° 49'; K = 121° 41'; k = 5.381 \text{ m; area} = 4.025 \text{ m}^2 \\ \text{OR } L = 149° 11'; K = 3° 19'; k = 0.366 \text{ m; area} = 0.071 \text{ m}^2 \end{bmatrix}$$

4. Use the cosine and sine rules to solve the following triangles PQR and find their areas.
   (a) $q = 12$ cm; $r = 16$ cm; $P = 54°$
   (b) $p = 56$ mm; $q = 38$ mm; $R = 64°$
   (c) $q = 3.25$ m; $r = 4.42$ m; $P = 105°$

   $$\begin{bmatrix} \text{(a) } p = 13.2 \text{ cm}; Q = 47° 21'; R = 78° 39'; \text{ area} = 77.7 \text{ cm}^2 \\ \text{(b) } r = 52.1 \text{ mm}; Q = 40° 58'; P = 75° 2'; \text{ area} = 956 \text{ mm}^2 \\ \text{(c) } p = 6.127 \text{ m}; Q = 30° 49'; R = 44° 11'; \text{ area} = 6.938 \text{ m}^2 \end{bmatrix}$$

5. Use the cosine and sine rules to solve the following triangles XYZ and find their areas.
   (a) $x = 10.0$ cm; $y = 8.0$ cm; $z = 7.0$ cm
   (b) $x = 2.4$ m; $y = 3.6$ m; $z = 1.5$ m
   (c) $x = 21$ mm; $y = 34$ mm; $z = 42$ mm

   $$\begin{bmatrix} \text{(a) } X = 83° 20'; Y = 52° 37'; Z = 44° 3'; \text{ area} = 27.8 \text{ cm}^2 \\ \text{(b) } X = 28° 57'; Y = 133° 26'; Z = 17° 37'; \text{ area} = 1.31 \text{ m}^2 \\ \text{(c) } X = 29° 46'; Y = 53° 31'; Z = 96° 43'; \text{ area} = 355 \text{ mm}^2 \end{bmatrix}$$

6. A ship P sails at a steady speed of 45 km/h in a direction of W 32° N (i.e. a bearing of 302°) from a port. At the same time another ship Q leaves the port at a steady speed of 35 km/h in a direction N 15° E (i.e. a bearing of 015°). Determine their distance apart after 4 hours.

   [193 km]

7. Two sides of an acute angled triangular plot of land are 52.0 m and 34.0 m respectively. If the area of the plot is 620 m² find (a) the length of fencing required to enclose the plot and (b) the angles of the triangular plot.

   [(a) 122.6 m; (b) 94° 49'; 40° 39'; 44° 32']

8  A jib crane is shown in *Fig 19*. If the tie rod PR is 8.0 long and PQ is 4.5 m long determine (a) the length of jib RQ and (b) the angle between the jib and the tie rod.
[(a) 11.4 m; (b) 17° 33′]

9  A building site is in the form of a quadrilateral as shown in *Fig 20*, and its area is 1296 m². Determine the length of the perimeter of the site.
[150.5 m]

10 Determine the length of members BF and EB in the roof truss shown in *Fig 21*.
[BF = 3.9 m; EB = 4.0 m]

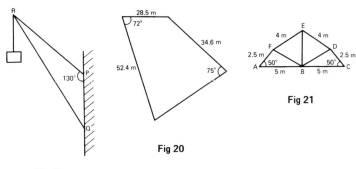

Fig 19  Fig 20  Fig 21

11 A laboratory 9.0 m wide has a span roof which slopes at 36° on one side and 44° on the other. Determine the lengths of the roof slopes.
[6.35 m; 5.37 m]

12 PQ and QR are the phasors representing the alternating currents in two branches of a circuit. Phasor PQ is 20.0 A and is horizontal. Phasor QR (which is joined to the end of PQ to form triangle PQR) is 14.0 A and is at an angle of 35° to the horizontal. Determine the resultant phasor PR and the angle it makes with phasor PQ.
[32.48 A; 14° 19′]

13 A vertical aerial AB 9.60 m high stands on ground which is inclined 12° to the horizontal. A stay connects the top of the aerial A to a point C on the ground 10.0 m downhill from B, the foot of the aerial. Determine (a) the length of the stay and (b) the angle the stay makes with the ground.
[(a) 15.23 m; (b) 38° 4′]

14 A vertical 35.0 cm by 35.0 cm ventilation shaft has an end covered by a plate that makes an angle of 21° 15′ with the horizontal. Determine the area of the plate.
[1314 cm²]

15 A chimney stack has a diameter of 1.5 m and passes through a roof that has a pitch of 36° 30′. Determine the area of the resulting void in the roof covering.
[2.20 m²]

16 Three forces acting on a fixed point are represented by the sides of a triangle of dimensions 7.2 cm, 9.6 cm and 11.0 cm. Determine the angles between the lines of action and the three forces.
[80° 25′; 59° 23′; 40° 12′]

17 A reciprocating engine mechanism is shown in *Fig 22*. The crank AB is 12.0 cm long and the connecting rod BC is 32.0 cm long. For the position shown determine the length of AC and the angle between the crank and the connecting rod.

[40.25 cm; 126° 3']

Fig 22

18 From *Fig 22*, determine how far C moves, correct to the nearest millimetre when angle CAB changes from 40° to 160°, B moving in an anticlockwise direction.

[19.8 cm]

19 A surveyor, standing W 25° S of a tower measures the angle of elevation of the top of the tower as 46° 30'. From a position E 23° S from the tower the elevation of the top is 37° 15'. Determine the height of the tower if the distance between the two observations is 75 m.

[36.2 m]

20 An aeroplane is sighted due east from a radar station at an elevation of 40° and a height of 8000 m and later at an elevation of 35° and height 5500 m in a direction E 70° S. If it is descending uniformly, find the angle of descent. Determine also the speed of the aeroplane in km/h if the time between the two observations is 45 s.

[13° 57'; 829.9 km/h]

# 3 Trigonometric graphs and the combination of waveforms

## A. MAIN POINTS CONCERNED WITH TRIGONOMETRIC GRAPHS AND COMBINATION OF WAVEFORMS

1  (i) A graph of $y = \sin A$ is shown by the broken line in *Fig 1* and is obtained by drawing up a table of values as in para 6 of chapter 1. A similar table may be produced for $y = \sin 2A$.

| $A°$    | 0     | 30    | 45   | 60    | 90  | 120    | 135   | 150    | 180 |
|---------|-------|-------|------|-------|-----|--------|-------|--------|-----|
| $2A$    | 0     | 60    | 90   | 120   | 180 | 240    | 270   | 300    | 360 |
| $\sin 2A$ | 0   | 0.866 | 1.0  | 0.866 | 0   | −0.866 | −1.0  | −0.866 | 0   |

| $A°$    | 210   | 225   | 240   | 270 | 300    | 315   | 330    | 360 |
|---------|-------|-------|-------|-----|--------|-------|--------|-----|
| $2A$    | 420   | 450   | 480   | 540 | 600    | 630   | 660    | 720 |
| $\sin 2A$ | 0.866 | 1.0 | 0.866 | 0   | −0.866 | −1.0  | −0.866 | 0   |

A graph of $y = \sin 2A$ is shown in *Fig 1*.

(ii) A graph of $y = \sin \frac{1}{2}A$ is shown in *Fig 2* using the following table of values.

| $A°$             | 0 | 30    | 60   | 90    | 120   | 150   | 180 | 210   |
|------------------|---|-------|------|-------|-------|-------|-----|-------|
| $\frac{1}{2}A$   | 0 | 15    | 30   | 45    | 60    | 75    | 90  | 105   |
| $\sin \frac{1}{2}A$ | 0 | 0.259 | 0.50 | 0.707 | 0.866 | 0.966 | 1.0 | 0.966 |

| $A°$             | 240   | 270   | 300  | 330   | 360 |
|------------------|-------|-------|------|-------|-----|
| $\frac{1}{2}A$   | 120   | 135   | 150  | 165   | 180 |
| $\sin \frac{1}{2}A$ | 0.866 | 0.707 | 0.50 | 0.259 | 0   |

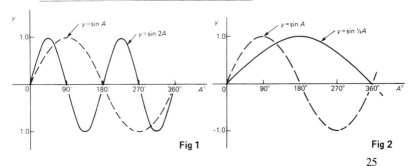

Fig 1                                     Fig 2

(iii) A graph of $y = \cos A$ is shown by the broken line in *Fig 3* and is obtained by drawing up a table of values as in para. 6 of chapter 1. A similar table may be produced for $y = \cos 2A$.

| $A°$ | 0 | 30 | 45 | 60 | 90 | 120 | 135 | 150 | 180 |
|---|---|---|---|---|---|---|---|---|---|
| $2A$ | 0 | 60 | 90 | 120 | 180 | 240 | 270 | 300 | 360 |
| $\cos 2A$ | 1.0 | 0.50 | 0 | −0.50 | −1.0 | −0.50 | 0 | 0.50 | 1.0 |

| $A°$ | 210 | 225 | 240 | 270 | 300 | 315 | 330 | 360 |
|---|---|---|---|---|---|---|---|---|
| $2A$ | 420 | 450 | 480 | 540 | 600 | 630 | 660 | 720 |
| $\cos 2A$ | 0.50 | 0 | −0.50 | −1.0 | −0.50 | 0 | 0.50 | 1.0 |

A graph of $y \cos 2A$ is shown in *Fig 3*.

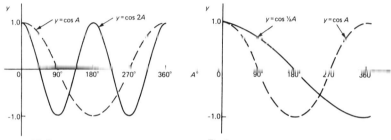

**Fig 3**                          **Fig 4**

(iv) A graph of $y = \cos \frac{1}{2}A$ is shown in *Fig 4* using the following table of values.

| $A°$ | 0 | 30 | 60 | 90 | 120 | 150 | 180 | 210 |
|---|---|---|---|---|---|---|---|---|
| $\frac{1}{2}A$ | 0 | 15 | 30 | 45 | 60 | 75 | 90 | 105 |
| $\cos \frac{1}{2}A$ | 1.0 | 0.966 | 0.866 | 0.707 | 0.50 | 0.259 | 0 | −0.259 |

| $A°$ | 240 | 270 | 300 | 330 | 360 |
|---|---|---|---|---|---|
| $\frac{1}{2}A$ | 120 | 135 | 150 | 165 | 180 |
| $\cos \frac{1}{2}A$ | −0.50 | −0.707 | −0.866 | −0.966 | −1.0 |

2 (i) Each of the graphs shown in *Figs 1 to 4* will repeat themselves as angle A increases and are thus called **periodic functions.**

(ii) $y = \sin A$ and $y = \cos A$ repeat themselves every 360° (or $2\pi$ radians); thus 360° is called the **period** of these waveforms. $y = \sin 2A$ and $y = \cos 2A$ repeat themselves every 180° (or $\pi$ radians); thus 180° is the period of these waveforms.

(iii) In general, if $y = \sin pA$ or $y = \cos pA$ (where $p$ is a constant) then the period of the waveform is $360°/p$ (or $2\pi/p$ rads).
Hence if $y = \sin 3A$ then the period is 360/3, i.e. 120°,
and if $y = \cos 4A$ then the period is 360/4, i.e. 90°.

3 **Amplitude** is the name given to the maximum or peak value of a sine wave. Each of the graphs shown in *Figs 1 to 4* have an amplitude of +1 (i.e. they oscillate between +1 and −1). However, if $y = 4 \sin A$, each of the values in the table is multiplied by 4 and the maximum value, and thus amplitude, is 4. Similarly, if $y = 5 \cos 2A$, the amplitude is 5 and the period is $360°/2$, i.e. $180°$.

4 **Lagging and leading angles**
   (i) A sine or cosine curve may not always start at $0°$. To show this a periodic function is represented by $y = \sin(A \pm \alpha)$ or $y = \cos(A \pm \alpha)$, where $\alpha$ is a phase displacement compared with $y = \sin A$ or $y = \cos A$.
   (ii) By drawing up a table of values, a graph of $y = \sin(A - 60°)$ may be plotted as shown in *Fig 5*. If $y = \sin A$ is assumed to start at $0°$ then $y = \sin(A - 60°)$ starts $60°$ later (i.e. has a zero value $60°$ later). Thus $y = \sin(A - 60°)$ is said to **lag** $y = \sin A$ by $60°$.
   (iii) By drawing up a table of values, a graph of $y = \cos(A + 45°)$ may be plotted as shown in *Fig 6*. If $y = \cos A$ is assumed to start at $0°$ then $y = \cos(A + 45°)$ starts $45°$ earlier (i.e. has a zero value $45°$ earlier). Thus $y = \cos(A + 45°)$ is said to **lead** $y = \cos A$ by $45°$.
   (iv) Generally, a graph of $y = \sin(A - \alpha)$ lags $y = \sin A$ by angle $\alpha$, and a graph of $y = \sin(A + \alpha)$ leads $y = \sin A$ by angle $\alpha$.
   (v) A cosine curve is the same shape as a sine curve but starts $90°$ earlier, i.e. leads by $90°$. Hence $\cos A = \sin(A + 90°)$.

5 **Graphs of $\sin^2 A$ and $\cos^2 A$**
   (i) A graph of $y = \sin^2 A$ is shown in *Fig 7* using the following table of values.

| $A°$ | 0 | 30 | 60 | 90 | 120 | 150 | 180 | 210 |
|---|---|---|---|---|---|---|---|---|
| $\sin A$ | 0 | 0.50 | 0.866 | 1.0 | 0.866 | 0.50 | 0 | −0.50 |
| $(\sin A)^2 = \sin^2 A$ | 0 | 0.25 | 0.75 | 1.0 | 0.75 | 0.25 | 0 | 0.25 |

| $A°$ | 240 | 270 | 300 | 330 | 360 |
|---|---|---|---|---|---|
| $\sin A$ | −0.866 | −1.0 | −0.866 | −0.50 | 0 |
| $(\sin A)^2 = \sin^2 A$ | 0.75 | 1.0 | 0.75 | 0.25 | 0 |

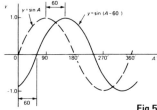

Fig 5

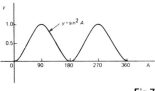

Fig 7

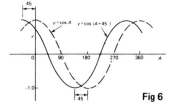

Fig 6

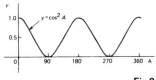

Fig 8

(ii) A graph of $y = \cos^2 A$ is shown in *Fig 8* using the following table of values.

| $A°$ | 0 | 30 | 60 | 90 | 120 | 150 | 180 | 210 |
|---|---|---|---|---|---|---|---|---|
| $\cos A$ | 1.0 | 0.866 | 0.50 | 0 | −0.50 | −0.866 | −1.0 | −0.866 |
| $(\cos A)^2 = \cos^2 A$ | 1.0 | 0.75 | 0.25 | 0 | 0.25 | 0.75 | 1.0 | 0.75 |

| $A°$ | 240 | 270 | 300 | 330 | 360 |
|---|---|---|---|---|---|
| $\cos A$ | −0.50 | 0 | 0.50 | 0.866 | 1.0 |
| $(\cos A)^2 = \cos^2 A$ | 0.25 | 0 | 0.25 | 0.75 | 1.0 |

(iii) $y = \sin^2 A$ and $y = \cos^2 A$ are both periodic functions of period $180°$ (or $\pi$ rads) and both contain only positive values. Thus a graph of $y = \sin^2 2A$ has a period $180°/2$, i.e. $90°$. Similarly, a graph of $y = 4\cos^2 3A$ has an amplitude of 4 and a period of $180/3$, i.e. $60°$.

## 6 Phasors

In *Fig 9*, let OR represent a vector that is free to rotate anticlockwise about 0 at a velocity of $\omega$ rads/s. A rotating vector is called a **phasor**. After a time $t$ seconds OR will have turned through an angle $\omega t$ radians (shown as angle TOR in *Fig 9*). If ST is constructed perpendicular to OR, then $\sin \omega t = \text{ST/OT}$, i.e. $\mathbf{ST = OT \sin \omega t}$.

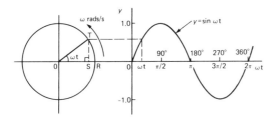

**Fig 9**

If all such vertical components are projected on to a graph of $y$ against $\omega t$, a sine wave results of amplitude OR (as shown in chapter 11 of *Mathematics I Checkbook*).

(ii) If phasor OR makes one revolution (i.e. $2\pi$ radians) in $T$ seconds, then the angular velocity, $\omega = 2\pi/T$ rads/s, from which, $T = 2\pi/\omega$ **seconds**. $T$ is known as the **periodic time**.

(iii) The number of complete cycles occurring per second is called the **frequency**, $f$.

$$\text{Frequency} = \frac{\text{number of cycles}}{\text{second}} = \frac{1}{T} = \frac{\omega}{2\pi} \text{ Hz, i.e. } f = \frac{\omega}{2\pi} \text{ Hz}.$$

Hence angular velocity, $\omega = 2\pi f$ rads/s.

(iv) Given a general sinusoidal periodic function $y = A \sin(\omega t \pm \alpha)$, then $A$ = amplitude, $\omega$ = angular velocity, $2\pi/\omega$ = periodic time, $T$, $\omega/2\pi$ = frequency, $f$, and $\alpha$ = angle of load or lag (compared with $y = A \sin \omega t$).

## 7 Combination of two periodic functions of the same frequency

There are a number of instances in engineering and science where waveforms combine and where it is required to determine the single phasor (called the resultant) which could replace two or more separate phasors. Uses are found in electrical alternating current theory, in mechanical vibrations, in the addition of forces and with sound waves. There are several methods of determining the resultant and two such methods are shown below.

(i) **Plotting the periodic functions graphically**

This may be achieved by sketching the separate functions on the same axes and then adding (or subtracting) ordinates at regular intervals (see *Problems 13 to 15*). Alternatively, a table of values may be drawn up before plotting the resultant waveforms (see *Problem 16*).

(ii) **Resolution of phasors by drawing or calculation**

The resultant of two periodic functions may be found from their relative positions when the time is zero. For example, if $y_1 = 4 \sin \omega t$ and $y_2 = 3 \sin (\omega t - \pi/3)$ then each may be represented as phasors as shown in *Fig 10*, $y_1$ being 4 units long and drawn horizontally and $y_2$ being 3 units long, lagging $y_1$ by $\pi/3$ radians or $60°$.

To determine the resultant of $y_1 + y_2$, $y_1$ is drawn horizontally as shown in

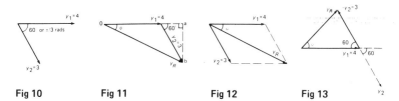

**Fig 10**       **Fig 11**       **Fig 12**       **Fig 13**

*Fig 11* and $y_2$ is joined to the end of $y_1$ at $60°$ to the horizontal. The resultant is given by $y_R$. This is the same as the diagonal of a parallelogram which is shown completed in *Fig 12*. Resultant $y_R$, in *Figs 11 and 12*, is determined either by:

(a) scaled drawing and measurement, or

(b) by use of the cosine rule (and then sine rule to calculate angle $\phi$), or

(c) by determining horizontal and vertical components of lengths oa and ab in *Fig 11*, and then using Pythagoras' theorem to calculate ob.

In this case, by calculation, $y_R = 6.083$ and angle $\phi = 25.28°$ or 0.441 rads. Thus the resultant may be expressed in sinusoidal form as

$y_R = 6.083 \sin (\omega t - 0.441)$. If the resultant phasor, $y_R = y_1 - y_2$ is required, then $y_2$ is still 3 units long but is drawn in the opposite direction, as shown in *Fig 13*, and $y_R$ is determined by measurement or calculation. (See *Problems 17 to 19*.)

## B. WORKED PROBLEMS ON TRIGONOMETRIC GRAPHS AND COMBINATION OF WAVEFORMS

*Problem 1* Sketch $y = \sin 3A$ between $A = 0°$ and $A = 360°$.

Amplitude = 1; Period = $\dfrac{360°}{3} = 120°$.

A sketch of $y = \sin 3A$ is shown in *Fig 14*.

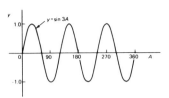

**Fig 14**

**Fig 17**

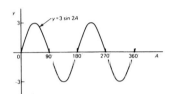

**Fig 15**

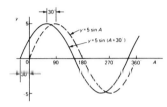

**Fig 18**

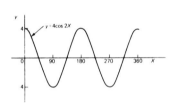

**Fig 16**

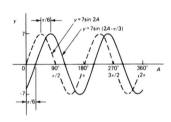

**Fig 19**

*Problem 2* Sketch $y = 3 \sin 2A$ from $A = 0$ to $A = 2\pi$ radians.

Amplitude = 3; Period = $2\pi/2 = \pi$ rads (or 180°).
A sketch of $y = 3 \sin 2A$ is shown in *Fig 15*.

*Problem 3* Sketch $y = 4 \cos 2x$ from $x = 0°$ to $x = 360°$

Amplitude = 4; Period = $360°/2 = 180°$.
A sketch of $y = 4 \cos 2x$ is shown in *Fig 16*.

*Problem 4* Sketch $y = 2 \sin \dfrac{3}{5} A$ over one cycle.

30

Amplitude = 2;  Period = $\dfrac{360°}{\frac{3}{5}} = \dfrac{360° \times 5}{3} = 600°$

A sketch of $y = 2 \sin \frac{3}{5} A$ is shown in *Fig 17*.

*Problem 5* Sketch $y = 5 \sin (A+30°)$ from $A = 0°$ to $A = 360°$.

Amplitude = 5;  Period = $\dfrac{360°}{1} = 360°$.

$5 \sin (A+30°)$ leads $5 \sin A$ by $30°$ (i.e. starts $30°$ earlier).
A sketch of $y = 5 \sin (A+30°)$ is shown in *Fig 18*.

*Problem 6* Sketch $y = 7 \sin (2A-\pi/3)$ over one cycle.

Amplitude = 7;  Period = $2\pi/2 = \pi$ radians.
In general, $y = \sin (pt-\alpha)$ lags $y = \sin pt$ by $\alpha/p$
Hence $7 \sin (2A-\pi/3)$ lags $7 \sin 2A$ by $\pi/\!\left(\dfrac{3}{2}\right)$, i.e. $\pi/6$ rads or $30°$.
A sketch of $y = 7 \sin (2A-\pi/3)$ is shown in *Fig 19*.

*Problem 7* Sketch $y = 2 \cos (\omega t-3\pi/10)$ over one cycle.

Amplitude = 2;  Period = $2\pi/\omega$ rads.
$2 \cos (\omega t-3\pi/10)$ lage $2 \cos \omega t$ by $3\pi/10\omega$ radians.
A sketch of $y = 2 \cos (\omega t-3\pi/10)$ is shown in *Fig 20*.

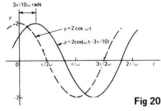

**Fig 20**

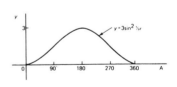

**Fig 21**

*Problem 8* Sketch $y = 3 \sin^2 \frac{1}{2} A$ in the range $0 \leqslant A \leqslant 360°$.

Amplitude = 3;  Period = $180°/\frac{1}{2} = 360°$ (see para. 5).
A sketch of $3 \sin^2 \frac{1}{2} A$ is shown in *Fig 21*.

31

*Problem 9* Sketch $y = 7 \cos^2 2A$ between $A = 0°$ and $A = 360°$.

Amplitude = 7; Period = $180°/2 = 90°$.
A sketch of $y = 7 \cos^2 2A$ is shown in *Fig 22*.

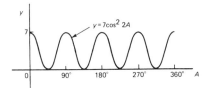

**Fig 22**

*Problem 10* An alternating current is given by $i = 30 \sin(100\pi t + 0.27)$ amperes. Find the amplitude, periodic time, frequency and phase angle (in degrees and minutes).

$i = 30 \sin(100\pi t + 0.27)$ A
Amplitude = **30 A**
Angular velocity $\omega = 100\pi$. Hence periodic time,

$$T = \frac{2\pi}{\omega} = \frac{2\pi}{100\pi} = \frac{1}{50}$$

= **0.02 s or 20 ms**

Frequency, $f = \frac{1}{T} = \frac{1}{0.02} =$ **50 Hz**

Phase angle, $\alpha = 0.27$ rads $= \left(0.27 \times \frac{180}{\pi}\right)° =$ **15° 28′ leading** $i = 30 \sin(100\pi t)$

*Problem 11* An oscillating mechanism has a maximum displacement of 2.5 m and a frequency of 60 Hz. At time $t = 0$ the displacement is 90 cm. Express the displacement in the general form $A \sin(\omega t \pm \alpha)$.

Amplitude = maximum displacement = 2.5 m.
Angular velocity, $\omega = 2\pi f = 2\pi(60) = 120\pi$ rads/s.
Hence displacement = $2.5 \sin(120\pi t + \alpha)$ m
When $t = 0$, displacement = 90 cm = 0.90 m.
Hence $\quad\quad\quad\quad 0.90 = 2.5 \sin(0 + \alpha)$
i.e. $\quad\quad\quad\quad \sin \alpha = \frac{0.90}{2.5} = 0.36$

Hence $\quad\quad\quad\quad \alpha = \arcsin 0.36 = 21° 6′ = 0.368$ rads.
**The displacement = 2.5 sin(120πt + 0.368) m**

*Problem 12* The instantaneous value of voltage in an ac circuit at any time $t$ seconds is given by $v = 340 \sin(50\pi t - 0.541)$ volts. Determine:
(a) the amplitude, periodic time, frequency and phase angle (in degrees),
(b) the value of the voltage when $t = 0$,
(c) the value of the voltage when $t = 10$ ms,
(d) the time when the voltage first reaches 200 V, and
(e) the time when the voltage is a maximum.
Sketch one cycle of the waveform.

(a) Amplitude = **340 V**
Angular velocity, $\omega = 50\pi$. Hence periodic time:
$$T = \frac{2\pi}{\omega} = \frac{2\pi}{50\pi} = \frac{1}{25} = 0.04 \text{ s or 40 ms}$$
Frequency $f = \frac{1}{T} = \frac{1}{0.04} = 25$ **Hz**
Phase angle = 0.541 rads = $\left(0.541 \times \frac{180}{\pi}\right)^{\circ} = 31°$ **lagging** $v = 340 \sin(50\pi t)$

(b) When $t = 0$, $v = 340 \sin(0 - 0.541) = 340 \sin(-31°) = $ **−175.1 V**
(c) When $t = 10$ ms then $v = 340 \sin(50\pi \frac{10}{10^3} - 0.541) = 340 \sin(1.0298)$
$= 340 \sin 59° = $ **291.4 volts.**
(d) When $v = 200$ volts then $200 = 340 \sin(50\pi t - 0.541)$
$$\frac{200}{340} = \sin(50\pi t - 0.541)$$
Hence $(50\pi t - 0.541) = \arcsin \frac{200}{340} = 36.03°$ or 0.6288 rads
$50\pi t = 0.6288 + 0.541 = 1.1698$
Hence time, $t = \frac{1.1698}{50\pi} = $ **7.447 ms**

(e) When the voltage is a maximum, $v = 340$ V
Hence $340 = 340 \sin(50\pi t - 0.541)$
$1 = \sin(50\pi t - 0.541)$
$50\pi t - 0.541 = \arcsin 1 = 90°$ or 1.5708 rads.
$50\pi t = 1.5708 + 0.541 = 2.1118$
Hence time, $t = \frac{2.1118}{50\pi} = $ **13.44 ms**

A sketch of $v = 340 \sin(50\pi t - 0.541)$ volts is shown in *Fig 23*.

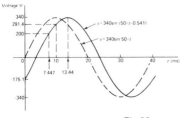

Fig 23

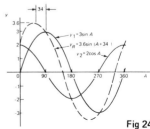

Fig 24

*Problem 13* Plot the graph of $y_1 = 3 \sin A$ from $A = 0°$ to $A = 360°$. On the same axes plot $y_2 = 2 \cos A$. By adding ordinates plot $y_R = 3 \sin A + 2 \cos A$ and obtain a sinusoidal expression for this resultant waveform.

$y_1 = 3 \sin A$ and $y_2 = 2 \cos A$ are shown plotted in *Fig 24*. Ordinates may be added at, say, 15° intervals. For example,

at   $0°$, $y_1 + y_2 = 0 + 2 = 2$,
at  $15°$, $y_1 + y_2 = 0.78 + 1.93 = 2.71$,
at $120°$, $y_1 + y_2 = 2.6 + -1 = 1.6$,
at $210°$, $y_1 + y_2 = -1.5 - 1.73 = -3.23$, and so on.

The resultant waveform, shown by the broken line has the same period, i.e. 360°, and thus the same frequency, as the single phasors. The maximum value, or amplitude, of the resultant is 3.6. The resultant waveform leads $y_1 = 3 \sin A$ by 34° or 0.593 rads. The sinusoidal expression for the resultant waveform is:

$y_R = 3.6 \sin (A + 34°)$ or $y_R = 3.6 \sin (A + 0.593)$

*Problem 14* Plot the graphs of $y_1 = 4 \sin \omega t$ and $y_2 = 3 \sin (\omega t - \pi/3)$ on the same axis, over one cycle. By adding ordinates at intervals plot $y_R = y_1 + y_2$ and obtain a sinusoidal expression for the resultant waveform.

$y_1 = 4 \sin \omega t$ and $y_2 = 3 \sin (\omega t - \pi/3)$ are shown plotted in *Fig 25*. Ordinates are added at 15° intervals and the resultant is shown by the broken line. The amplitude of the resultant is 6.1 and it lags $y_1$ by 25° or 0.436 rads. Hence the sinusoidal expression for the resultant waveform is:
$y_R = 6.1 \sin (\omega t - 0.436)$.

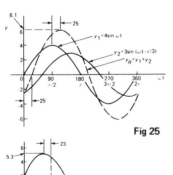

Fig 25

Fig 27

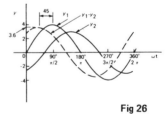

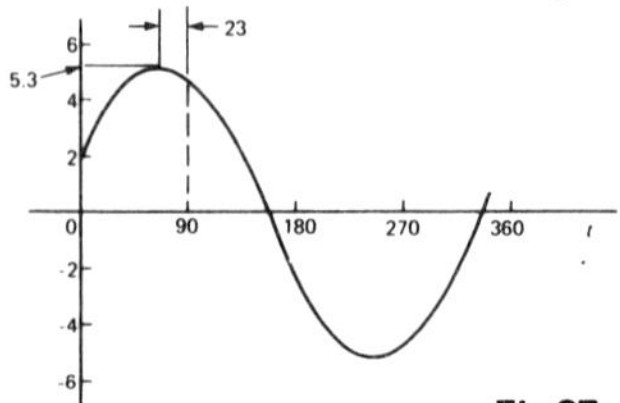

Fig 26

*Problem 15* Determine a sinusoidal expression for $y_1 - y_2$ when $y_1 = 4 \sin \omega t$ and $y_2 = 3 \sin (\omega t - \pi/3)$.

$y_1$ and $y_2$ are shown plotted in *Fig 26*. At 15° intervals $y_2$ is subtracted from $y_1$. For example:

at  0°, $y_1 - y_2 = 0 - (-2.6) = +2.6$,
at  30°, $y_1 - y_2 = 2 - (-1.5) = +3.5$
at 150°, $y_1 - y_2 = 2 - 3 = -1$, and so on.

The amplitude, or peak value of the resultant (shown by the broken line), is 3.6 and it leads $y_1$ by 45° or 0.79 rads.
Hence $y_1 - y_2 = 3.6 \sin (\omega t + 0.79)$.

*Problem 16* Draw a graph to represent current $i = 2.4 \sin t + 3.2 \sin (t+40°)$ amperes and express $i$ in the general form $i = A \sin (t \pm \alpha)$.

A table of values may be drawn up as shown below.

| $t°$ | 0 | 30 | 60 | 90 | 120 | 150 | 180 | 210 |
|---|---|---|---|---|---|---|---|---|
| $\sin t$ | 0 | 0.5 | 0.866 | 1.0 | 0.866 | 0.5 | 0 | −0.5 |
| 2.4 $\sin t$ | 0 | 1.2 | 2.08 | 2.4 | 2.08 | 1.2 | 0 | −1.2 |
| $(t+40°)$ | 40 | 70 | 100 | 130 | 160 | 190 | 220 | 250 |
| $\sin (t+40°)$ | 0.64 | 0.94 | 0.98 | 0.77 | 0.34 | −0.17 | −0.64 | −0.94 |
| 3.2 $\sin (t+40°)$ | 2.05 | 3.01 | 3.14 | 2.46 | 1.09 | −0.54 | −2.05 | −3.01 |
| $i = 2.4 \sin t +$ 3.2 $\sin (t+40°)$ | 2.05 | 4.21 | 5.22 | 4.86 | 3.17 | 0.66 | −2.05 | −4.21 |

| $t°$ | 240 | 270 | 300 | 330 | 360 |
|---|---|---|---|---|---|
| $\sin t$ | −0.866 | −1.0 | −0.866 | −0.5 | 0 |
| 2.4 $\sin t$ | −2.08 | −2.4 | −2.08 | −1.2 | 0 |
| $(t+40°)$ | 280 | 310 | 340 | 370 | 400 |
| $\sin (t+40°)$ | −0.98 | −0.77 | −0.34 | 0.17 | 0.64 |
| 3.2 $\sin (t+40°)$ | −3.14 | −2.46 | −1.09 | 0.54 | 2.05 |
| $i = 2.4 \sin t +$ 3.2 $\sin (t+40°)$ | −5.22 | −4.48 | −3.17 | −0.66 | 2.05 |

A graph of $i = 2.4 \sin t + 3.2 \sin (t+40°)$ is shown in *Fig 27*.
The amplitude is 5.3 amperes and $i$ is 23°, i.e. 0.40 rads, ahead of a sine wave starting at 0°.
**Hence $i = 5.3 \sin (t+0.40)$ amperes.**

*Problem 17* Given $y_1 = 2 \sin \omega t$ and $y_2 = 3 \sin(\omega t + \pi/4)$, obtain an expression of the resultant $y_R = y_1 + y_2$, (a) by drawing and (b) by calculation.

(a) When time $t = 0$ the position of phasors $y_1$ and $y_2$ are as shown in *Fig 28(a)*. To obtain the resultant, $y_1$ is drawn horizontally, 2 units long, $y_2$ is drawn 3 units long at an angle of $\pi/4$ rads or $45°$ and joined to the end of $y_1$ as shown in *Fig 28(b)*. $y_R$ is measured as 4.6 units long and angle $\phi$ is measured as $27°$ or 0.47 rads. Alternatively, $y_R$ is the diagonal of the parallelogram formed as shown in *Fig 28(c)*.
Hence, by drawing, $y_R = 4.6 \sin(\omega t + 0.47)$

(b) From *Fig 28(b)*, and using the cosine rule:

$$y_R^2 = 2^2 + 3^2 - [2(2)(3) \cos 135°] = 4 + 9 - [-8.485] = 21.49$$

Hence $y_R = \sqrt{(21.49)} = 4.64$

Using the sine rule: $\dfrac{3}{\sin \phi} = \dfrac{4.64}{\sin 135°}$, from which, $\sin \phi = \dfrac{3 \sin 135°}{4.64}$
$= 0.4572$

Hence $\phi = \arcsin 0.4572 = 27°12'$ or 0.475 rads.
By calculation, $y_R = 4.64 \sin(\omega t + 0.475)$

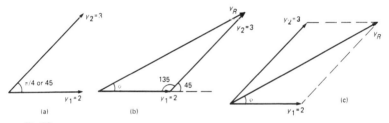

**Fig 28**

*Problem 18* Two alternating voltages are given by $v_1 = 15 \sin \omega t$ volts and $v_2 = 25 \sin(\omega t - \pi/6)$ volts. Determine a sinusoidal expression for the resultant $v_R = v_1 + v_2$ by finding horizontal and vertical components.

The relative positions of $v_1$ and $v_2$ at time $t = 0$ are shown in *Fig 29(a)* and the phasor diagram is shown in *Fig 29(b)*.
The horizontal component of $v_R$ = oa + ab
$= 15 + 25 \cos 30° = 36.65$ V.
The vertical component of $v_R$ = bc $= 25 \sin 30° = 12.50$ V.

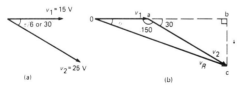

**Fig 29**

Hence $v_R$ (= oc) = $\sqrt{[(36.65)^2 + (12.50)^2]}$ by Pythagoras' theorem
= 38.72 volts.

$$\tan \phi = \frac{bc}{ob} = \frac{12.50}{36.65} = 0.3411,$$

from which, $\phi$ = arctan 0.3411 = 18° 50' or 0.329 radians.
**Hence $v_R = v_1 + v_2 = 38.72 \sin(\omega t - 0.329)$ V**

*Problem 19* For the voltages in *Problem 18*, determine the resultant $v_R = v_1 - v_2$.

To find the resultant $v_R = v_1 - v_2$, the phasor $v_2$ of *Fig 29(b)* is reversed in direction as shown in *Fig 30*.
Using the cosine rule: $v_R^2 = 15^2 + 25^2 - 2(15)(25) \cos 30°$
= 225 + 625 - 649.5 = 200.5.
$v_R = \sqrt{(200.5)} = 14.16$ volts.
Using the sine rule: $\frac{25}{\sin \phi} = \frac{14.16}{\sin 30°}$, from which, $\sin \phi = \frac{25 \sin 30°}{14.16} = 0.8828.$

Hence $\phi$ = arcsin 0.8828 = 61.98° or 118.02°
From *Fig 30*, $\phi$ is obtuse,
hence $\phi$ = 118.02°
or 2.06 radians.
**Hence $v_R = v_1 - v_2 = 14.16 \sin(\omega t + 2.06)$ V**

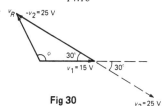

**Fig 30**

## C. FURTHER PROBLEMS ON TRIGONOMETRICAL GRAPHS AND COMBINATION OF WAVEFORMS

In *Problems 1 to 3* state the amplitude and period of the waveform and sketch the curve between 0° and 360°

1  (a) $y = \cos 3A$; (b) $y = 2 \sin \frac{5x}{2}$; (c) $y = 3 \sin 4t$
[(a) 1, 120°; (b) 2, 144°; (c) 3, 90°]

2  (a) $y = 3 \cos \frac{\theta}{2}$; (b) $y = \frac{7}{2} \sin \frac{3x}{8}$; (c) $y = 6 \sin(t - 45°)$
[(a) 3, 720°; (b) $\frac{7}{2}$, 960°; (c) 6, 360°]

3  (a) $y = 4 \cos(2\theta + 30°)$; (b) $y = 2 \sin^2 2t$; (c) $y = 5 \cos^2 \frac{3}{2}\theta$.
[(a) 4, 180°; (b) 2, 90°; (c) 5, 120°]

In *Problems 4 to 6* state the amplitude, periodic time, frequency and phase angle (in degrees).

4  $v = 200 \sin(200\pi t + 0.29)$ V.
[200 V; 10 ms; 100 Hz; 16° 37' leading]

5  $i = 32 \sin(400\pi t - 0.42)$ A.
[32 A; 5 ms; 200 Hz; 24° 4' lagging]

6  $x = 5 \sin(314.2t + 0.33)$ cm.
[5 cm; 20 ms; 50 Hz. 18° 54' leading]

7. A sinusoidal voltage has a maximum value of 120 V and a frequency of 50 Hz. At time $t = 0$, the voltage is (a) zero, and (b) 50 V. Express the instantaneous voltage $v$ in the form $v = A \sin(\omega t \pm \alpha)$.

[(a) $v = 120 \sin 100\pi t$ volts; (b) $v = 120 \sin(100\pi t + 0.43)$ volts]

8. An alternating current has a periodic time of 25 ms and a maximum value of 20 A. When time $t = 0$, current $i = -10$ amperes. Express the current $i$ in the form $i = A \sin(\omega t \pm \alpha)$.

$[i = 20 \sin(80\pi t - \frac{\pi}{6})$ amperes]

9. The current in an ac circuit at any time $t$ seconds is given by:

$i = 5 \sin(100\pi t - 0.432)$ amperes.

Determine (a) the amplitude, periodic time, frequency and phase angle (in degrees), (b) the value of current at $t = 0$; (c) the value of current at $t = 8$ ms; (d) the time when the current is first a maximum; (e) the time when the current first reaches 3 A. Sketch one cycle of the waveform showing relevant points.

[(a) 5 A; 20 ms, 50 Hz, 24° 45' lagging; (b) −2.093 A; (c) 4.363 A; (d) 6.375 ms; (e) 3.423 ms.]

10. Plot the graph of $y = 2 \sin A$ from $A = 0°$ to $A = 360°$. On the same axes plot $y = 4 \cos A$. By adding ordinates at intervals plot $y = 2 \sin A + 4 \cos A$ and obtain a sinusoidal expression for the waveform.

[$4.5 \sin(A + 63° 26')$]

11. Two alternating voltages are given by $v_1 = 10 \sin \omega t$ volts and $v_2 = 14 \sin(\omega t + \pi/3)$ volts. By plotting $v_1$ and $v_2$ on the same axes over one cycle obtain a sinusoidal expression for (a) $v_1 + v_2$; (b) $v_1 - v_2$.

[(a) $20.9 \sin(\omega t + 0.62)$ volts; (b) $12.5 \sin(\omega t - 1.33)$ volts]

12. Draw up a table of values for the waveform $y = 5 \sin(\omega t + \pi/4) + 3 \sin(\omega t - \pi/3)$ and plot a graph of $y$ against $\omega t$. Express $y$ in the form $y = A \sin(\omega t \pm \alpha)$.

[$5.1 \sin(\omega t + 0.184)$]

In *Problems 13 and 14*, express the combination of periodic functions in the form $A \sin(\omega t \pm \alpha)$ either by drawing or by calculation.

13. (a) $12 \sin \omega t + 5 \cos \omega t$
    (b) $7 \sin \omega t + 5 \sin(\omega t + \frac{\pi}{4})$
    (c) $6 \sin \omega t + 3 \sin(\omega t - \frac{\pi}{6})$

    [(a) $13 \sin(\omega t + 0.395)$
    (b) $11.11 \sin(\omega t + 0.324)$
    (c) $8.73 \sin(\omega t - 0.173)$]

14. (a) $i = 25 \sin \omega t - 15 \sin(\omega t + \frac{\pi}{3})$
    (b) $v = 8 \sin \omega t - 5 \sin(\omega t - \frac{\pi}{4})$
    (c) $x = 9 \sin(\omega t + \frac{\pi}{3}) - 7 \sin(\omega t - \frac{3\pi}{8})$

    [(a) $i = 21.79 \sin(\omega t - 0.639)$
    (b) $v = 5.695 \sin(\omega t + 0.670)$
    (c) $x = 14.38 \sin(\omega t + 1.444)$]

# 4 Trigonometric identities and the solution of equations

## A MAIN POINTS CONCERNED WITH TRIGONOMETRIC IDENTITIES AND THE SOLUTION OF TRIGONOMETRIC EQUATIONS

1   A **trigonometric identity** is an expression that is true for all values of the unknown variable.

$\tan \theta = \dfrac{\sin \theta}{\cos \theta}$, $\cot \theta = \dfrac{\cos \theta}{\sin \theta}$, $\sec \theta = \dfrac{1}{\cos \theta}$, $\csc \theta = \dfrac{1}{\sin \theta}$ and $\cot \theta = \dfrac{1}{\tan \theta}$

are examples of trigonometric identities (see chapter 1).

2   Applying Pythagoras' theorem to the right-angled triangle shown in *Fig 1* gives:

$a^2 + b^2 = c^2$                (1)

Dividing each term of equation (1) by $c^2$ gives:

$\dfrac{a^2}{c^2} + \dfrac{b^2}{c^2} = \dfrac{c^2}{c^2}$,   i.e.   $\left(\dfrac{a}{c}\right)^2 + \left(\dfrac{b}{c}\right)^2 = 1$

$(\cos \theta)^2 + (\sin \theta)^2 = 1$

Hence $\cos^2 \theta + \sin^2 \theta = 1$      (2)

Dividing each term of equation (1) by $a^2$ gives:

$\dfrac{a^2}{a^2} + \dfrac{b^2}{a^2} = \dfrac{c^2}{a^2}$   i.e.   $1 + \left(\dfrac{b}{a}\right)^2 = \left(\dfrac{c}{a}\right)^2$

Hence $1 + \tan^2 \theta = \sec^2 \theta$      (3)

Dividing each term of equation (1) by $b^2$ gives:

$\dfrac{a^2}{b^2} + \dfrac{b^2}{b^2} = \dfrac{c^2}{b^2}$,   i.e.   $\left(\dfrac{a}{b}\right)^2 + 1 = \left(\dfrac{c}{b}\right)^2$

Hence $\cot^2 \theta + 1 = \csc^2 \theta$      (4)

**Fig 1**

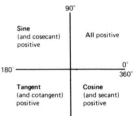

**Fig 2**

Equations (2), (3) and (4) are three further examples of trigonometric identities. For the proof of further trigonometric identities, see *Problems 1 to 6*.

3   Equations which contain trigonometric ratios are called **trigonometric equations**. There are usually an infinite number of solutions to such equations: however, solutions are often restricted to those between $0°$ and $360°$. A knowledge of angles of any magnitude is essential in the solution of trigonometric equations (see chapter 1) and calculators cannot be relied upon to give all the solutions. *Fig 2* shows a summary for angles of any magnitude.

4   **Equations of the type $a \sin^2 A + b \sin A + c = 0$**
   (i) When $a = 0$, $b \sin A + c = 0$

   Hence $\sin A = \dfrac{-c}{b}$   and   $A = \arcsin\left(\dfrac{-c}{b}\right)$

39

There are two values of $A$ between $0°$ and $360°$ which satisfy such an equation, provided $-1 \leq \frac{c}{b} \leq 1$, (see *Problems 7 to 10*).

(ii) **When** $b = 0$, $a \sin^2 A + c = 0$

Hence $\sin^2 A = \frac{-c}{a}$

$$\sin A = \sqrt{\left(\frac{-c}{a}\right)}$$

and $\quad A = \arcsin \sqrt{\left(\frac{-c}{a}\right)}$

If either '$a$' or '$c$' is a negative number, then the value within the square root sign is positive. Since when a square root is taken there is a positive and negative answer there are four values of $A$ between $0°$ and $360°$ which satisfy such an equation, provided $-1 \leq \frac{c}{a} \leq 1$. (See *Problems 11 to 13*.)

(iii) **When $a$, $b$ and $c$ are all non-zero:**
$a \sin^2 A + b \sin A + c = 0$ is a quadratic equation in which the unknown is $\sin A$. The solution of a quadratic equation is obtained either by factorising (if possible) or by using the quadratic formula:

$\sin A = \frac{-b \pm \sqrt{(b^2 - 4ac)}}{2a}$ (see *Problems 14 to 16*).

(iv) Often the trigonometric identities $\cos^2 A + \sin^2 A = 1$, $1 + \tan^2 A = \sec^2 A$ and $\cot^2 A + 1 = \csc^2 A$ need to be used to reduce equations to one of the above forms (see *Problems 17 to 20*).

## B. WORKED PROBLEMS ON TRIGONOMETRIC IDENTITIES AND THE SOLUTION OF TRIGONOMETRIC EQUATIONS

*Problem 1* Prove the identity $\sin^2 \theta \cot \theta \sec \theta = \sin \theta$

With trigonometric identities it is necessary to start with the left hand side (LHS) and attempt to make it equal to the right hand side (RHS) or vice versa. It is often useful to change all of the trigonometric ratios into sines and cosines where possible. Thus

LHS = $\sin^2 \theta \cot \theta \sec \theta = \sin^2 \theta \left(\frac{\cos \theta}{\sin \theta}\right) \left(\frac{1}{\cos \theta}\right) = \sin \theta$ (by cancelling) = RHS

*Problem 2* Prove that $\dfrac{\tan x + \sec x}{\sec x \left(1 + \dfrac{\tan x}{\sec x}\right)} = 1$.

$$\text{LHS} = \frac{\tan x + \sec x}{\sec x \left(1 + \frac{\tan x}{\sec x}\right)} = \frac{\frac{\sin x}{\cos x} + \frac{1}{\cos x}}{\left(\frac{1}{\cos x}\right)\left(1 + \frac{\frac{\sin x}{\cos x}}{\frac{1}{\cos x}}\right)} = \frac{\frac{\sin x + 1}{\cos x}}{\left(\frac{1}{\cos x}\right)\left[1 + \left(\frac{\sin x}{\cos x}\right)\left(\frac{\cos x}{1}\right)\right]}$$

$$= \frac{\frac{\sin x + 1}{\cos x}}{\frac{1}{\cos x}[1 + \sin x]} = \left(\frac{\sin x + 1}{\cos x}\right)\left(\frac{\cos x}{1 + \sin x}\right) = 1 \text{ (by cancelling)} = \text{RHS}$$

*Problem 3* Prove that $\frac{1+\cot\theta}{1+\tan\theta} = \cot\theta$.

$$\text{LHS} = \frac{1+\cot\theta}{1+\tan\theta} = \frac{1 + \frac{\cos\theta}{\sin\theta}}{1 + \frac{\sin\theta}{\cos\theta}} = \frac{\frac{\sin\theta+\cos\theta}{\sin\theta}}{\frac{\cos\theta+\sin\theta}{\cos\theta}} = \left(\frac{\sin\theta+\cos\theta}{\sin\theta}\right)\left(\frac{\cos\theta}{\cos\theta+\sin\theta}\right)$$

$$= \frac{\cos\theta}{\sin\theta} = \cot\theta = \text{RHS}.$$

*Problem 4* Show that $\cos^2\theta - \sin^2\theta = 1 - 2\sin^2\theta$.

From equation (2), para. 2, $\cos^2\theta + \sin^2\theta = 1$, from which, $\cos^2\theta = 1 - \sin^2\theta$.
Hence, LHS $= \cos^2\theta - \sin^2\theta = (1-\sin^2\theta) - \sin^2\theta = 1 - \sin^2\theta - \sin^2\theta$
$= 1 - 2\sin^2\theta = \text{RHS}$

*Problem 5* Prove that $\sqrt{\left(\frac{1-\sin x}{1+\sin x}\right)} = \sec x - \tan x$.

$$\text{LHS} = \sqrt{\left(\frac{1-\sin x}{1+\sin x}\right)} = \sqrt{\left\{\frac{(1-\sin x)(1-\sin x)}{(1+\sin x)(1-\sin x)}\right\}} = \sqrt{\left\{\frac{(1-\sin x)^2}{(1-\sin^2 x)}\right\}}.$$

Since $\cos^2 x + \sin^2 x = 1$ then $1 - \sin^2 x = \cos^2 x$,

$$\text{LHS} = \sqrt{\left\{\frac{(1-\sin x)^2}{(1-\sin^2 x)}\right\}} = \sqrt{\left\{\frac{(1-\sin x)^2}{\cos^2 x}\right\}} = \frac{1-\sin x}{\cos x} = \frac{1}{\cos x} - \frac{\sin x}{\cos x}$$

$= \sec x - \tan x = \text{RHS}$

*Problem 6* Use tables to show that the identity $1 + \cot^2\theta = \text{cosec}^2\theta$ is true when $\theta = 223°$.

$\cot 223° = +\cot 43° = +1.0724$. $\cot^2 223° = (+1.0724)^2 = 1.1500$
Hence LHS $= 1 + \cot^2\theta = 1 + 1.1500 = 2.1500$
$\text{cosec } 223° = -\text{cosec } 43° = -1.4663$. $\text{cosec}^2 223° = (-1.4663)^2 = 2.1500 = \text{RHS}$
Hence, the identity $1 + \cot^2\theta = \text{cosec}^2\theta$ is true when $\theta = 223°$

*Problem 7* Solve the trigonometric equation $5 \sin \theta + 3 = 0$ for values of $\theta$ from $0°$ to $360°$.

$5 \sin \theta + 3 = 0$, from which $\sin \theta = \dfrac{-3}{5} = -0.6000$. Hence $\theta = \arcsin(-0.6000)$.

Sine is negative in the third and fourth quadrants (see *Fig 3*).
The acute angle $\arcsin(0.6000) = 36° 52'$ (shown as $\alpha$ in *Fig 3(b)*).
Hence $\theta = 180° + 36° 52'$, i.e. **216° 52'**  or  $\theta = 360° - 36° 52'$, i.e. **323° 8'**.

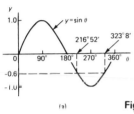 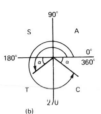

**Fig 3**

*Problem 8* Solve $1.5 \tan x - 1.8 = 0$ for $0° \leqslant x \leqslant 360°$.

$1.5 \tan x - 1.8 = 0$, from which $\tan x = \dfrac{1.8}{1.5} = 1.2000$. Hence $x = \arctan 1.2000$.

Tangent is positive in the first and third quadrants, (see *Fig 4*).
The acute angle $\arctan 1.2000 = 50° 12'$.
Hence $x = \mathbf{50° \ 12'}$ or $180° + 50° \ 12' = \mathbf{230° \ 12'}$.

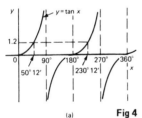 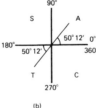 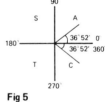

**Fig 4**           **Fig 5**

*Problem 9* Solve $4 \sec t = 5$ for values of $t$ between $0°$ and $360°$.

$4 \sec t = 5$, from which $\sec t = 5/4 = 1.2500$.
Hence $t = \text{arcsec } 1.2500$.

Secant $= \dfrac{1}{\text{cosine}}$ is positive in the first and fourth quadrants, (see *Fig 5*).

The acute angle $\text{arcsec } 1.2500 = 36° 52'$.
Hence $t = \mathbf{36° \ 52'}$ or $360° - 36° \ 52' = \mathbf{323° \ 8'}$.

*Problem 10* Solve $3.2(\cot\theta - 1) = -12$ for values of $\theta$ between $0°$ and $360°$.

$3.2(\cot\theta - 1) = -12$.

Hence $\cot\theta - 1 = \dfrac{-12}{3.2} = -3.7500$

$\cot\theta = -3.7500 + 1 = -2.7500$
$\theta = \text{arccot}(-2.7500)$

Cotangent $= \dfrac{1}{\text{tangent}}$ is negative in the second and fourth quadrants (see *Fig 6*).

The acute angle arccot $(2.7500) = 19°\ 59'$.
Hence $\theta = 180° - 19°\ 59' = 160°\ 1'$
or $\theta = 360° - 19°\ 59' = 340°\ 1'$.

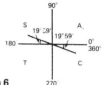

**Fig 6**

*Problem 11* Solve $2 - 4\cos^2 A = 0$ for values of $A$ in the range $0° \leqslant A \leqslant 360°$.

$2 - 4\cos^2 A = 0$, from which $\cos^2 A = 2/4 = 0.5000$.
Hence $\cos A = \sqrt{(0.5000)} = \pm 0.7071$ and $A = \arccos(\pm 0.7071)$
Cosine is positive in quadrants one and four and negative in quadrants two and three. Thus in this case there are four solutions, one in each quadrant (see *Fig 7*).
The acute angle $\arccos 0.7071 = 45°$. Hence $A = \mathbf{45°, 135°, 225°}$ or $\mathbf{315°}$.

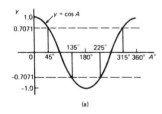

**Fig 7**

*Problem 12* Solve $0.75\sec^2 x - 1.2 = 0$ for values of $x$ between $0°$ and $360°$.

$0.75\sec^2 x - 1.2 = 0$, from which, $\sec^2 x = 1.2/0.75 = 1.6000$.
Hence $\sec x = \sqrt{(1.6000)} = \pm 1.2649$
and $x = \text{arcsec}(\pm 1.2649)$.
There are four solutions, one in each quadrant.
The acute angle $\text{arcsec}\ 1.2649 = 37°\ 46'$.
Hence $x = \mathbf{37°\ 46', 142°\ 14', 217°\ 46'}$ or $\mathbf{322°\ 14'}$.

*Problem 13* Solve $\frac{1}{2}\cot^2 y = 1.3$ for $0° \leq y \leq 360°$.

$\frac{1}{2}\cot^2 y = 1.3$, from which, $\cot^2 y = 2(1.3) = 2.6$.
Hence $\cot y = \sqrt{2.6} = \pm 1.6125$
and $y = \text{arccot}(\pm 1.6125)$.
There are four solutions, one in each quadrant.
The acute angle arccot $1.6125 = 31° \ 48'$.
Hence $y = \mathbf{31°\ 48', 148°\ 12', 211°\ 48'}$ or $\mathbf{328°\ 12'}$.

*Problem 14* Solve the *equation* $8 \sin^2 \theta + 2 \sin \theta - 1 = 0$, for all values of $\theta$ between 0° and 360°.

Factorising $8 \sin^2 \theta + 2 \sin \theta - 1 = 0$ gives $(4 \sin \theta - 1)(2 \sin \theta + 1) = 0$.

Hence $4 \sin \theta - 1 = 0$, from which, $\sin \theta = \frac{1}{4} = 0.2500$

or $2 \sin \theta + 1 = 0$, from which, $\sin \theta = -\frac{1}{2} = -0.5000$.

$\theta = \arcsin 0.2500 = 14° \ 29'$ or $165° \ 31'$, since sine is positive in the first and second quadrants,

or $\theta = \arcsin(-0.5000) = 210°$ or $330°$, since sine is negative in the third and fourth quadrants.

Hence $\theta = \mathbf{14°\ 29', 165°\ 31', 210°}$ or $\mathbf{330°}$.

*Problem 15* Solve $6 \cos^2 \theta + 5 \cos \theta - 6 = 0$ for values of $\theta$ from 0° to 360°.

Factorising $6 \cos^2 \theta + 5 \cos \theta - 6 = 0$ gives $(3 \cos \theta - 2)(2 \cos \theta + 3) = 0$.

Hence $3 \cos \theta - 2 = 0$, from which, $\cos \theta = \frac{2}{3} = 0.6667$

or $2 \cos \theta + 3 = 0$, from which, $\cos \theta = -\frac{3}{2} = -1.5000$.

The minimum value of a cosine is $-1$, hence the latter expression has no solution and is thus neglected.
Hence $\theta = \arccos 0.6667 = \mathbf{48°\ 11'}$ or $\mathbf{311°\ 49'}$, since cosine is positive to the first and fourth quadrants.

*Problem 16* Solve $9 \tan^2 x + 16 = 24 \tan x$ in the range $0° \leq x \leq 360°$.

Rearranging gives $9 \tan^2 x - 24 \tan x + 16 = 0$
and factorising gives $(3 \tan x - 4)(3 \tan x - 4) = 0$
i.e. $(3 \tan x - 4)^2 = 0$.

44

Hence $\tan x = \frac{4}{3} = 1.3333$ and $x = \arctan 1.3333 = $ **53° 8′ or 233° 8′**, since tangent is positive in the first and third quadrants.

*Problem 17* Solve $5\cos^2 t + 3\sin t - 3 = 0$ for values of $t$ from $0°$ to $360°$.

Since $\cos^2 t + \sin^2 t = 1$, $\cos^2 t = 1 - \sin^2 t$.
Substituting for $\cos^2 t$ in $5\cos^2 t + 3\sin t - 3 = 0$
gives $5(1 - \sin^2 t) + 3\sin t - 3 = 0$
$\quad\quad 5 - 5\sin^2 t + 3\sin t - 3 = 0$
$\quad\quad -5\sin^2 t + 3\sin t + 2 = 0$
$\quad\quad 5\sin^2 t - 3\sin t - 2 = 0$
Factorising gives $(5\sin t + 2)(\sin t - 1) = 0$
Hence $5\sin t + 2 = 0$, from which, $\sin t = -\frac{2}{5} = -0.4000$
or $\quad \sin t - 1 = 0$, from which, $\sin t = 1$.
$t = \arcsin(-0.4000) = 203° 35'$ or $336° 25'$, since sine is negative in the third and fourth quadrants, or $t = \arcsin 1 = 90°$.
Hence $t = $ **90°, 203° 35′ or 336° 25′**, as shown in *Fig 8*.

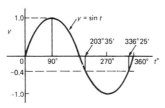

**Fig 8**

*Problem 18* Solve $18\sec^2 A - 3\tan A = 21$ for values of $A$ between $0°$ and $360°$.

$1 + \tan^2 A = \sec^2 A$ from para 2.
Substituting for $\sec^2 A$ in $18\sec^2 A - 3\tan A = 21$ gives $18(1 + \tan^2 A) - 3\tan A = 21$
i.e., $18 + 18\tan^2 A - 3\tan A - 21 = 0$
$\quad\quad 18\tan^2 A - 3\tan A - 3 = 0$
Factorising gives $(6\tan A - 3)(3\tan A + 1) = 0$
Hence $6\tan A - 3 = 0$, from which, $\tan A = \frac{3}{6} = 0.5000$
or $3\tan A + 1 = 0$, from which, $\tan A = -\frac{1}{3} = -0.3333$
Thus $A = \arctan(0.5000) = 26° 34'$ or $206° 34'$, since tangent is positive in the first and third quadrants,
or $\quad A = \arctan(-0.3333) = 161° 34'$ or $341° 34'$, since tangent is negative in the second and fourth quadrants.
Hence $A = $ **26° 34′, 161° 34′, 206° 34′ or 341° 34′**.

*Problem 19* Solve $3\operatorname{cosec}^2 \theta - 5 = 4\cot\theta$ in the range $0 \leq \theta \leq 360°$.

$\cot^2\theta + 1 = \operatorname{cosec}^2\theta$, from para. 2.
Substituting for $\operatorname{cosec}^2\theta$ in $3\operatorname{cosec}^2\theta - 5 = 4\cot\theta$
gives $\qquad 3(\cot^2\theta + 1) - 5 = 4\cot\theta$
$\qquad\qquad 3\cot^2\theta + 3 - 5 = 4\cot\theta$
$\qquad\qquad 3\cot^2\theta - 4\cot\theta - 2 = 0$.

Since the left hand side does not factorise the quadratic formula is used.

Thus $\cot\theta = \dfrac{-(-4) \pm \sqrt{[(-4)^2 - 4(3)(-2)]}}{2(3)} = \dfrac{4 \pm \sqrt{(16+24)}}{6}$

$= \dfrac{4 \pm \sqrt{40}}{6} = \dfrac{10.3246}{6}$ or $\dfrac{-2.3246}{6}$

Hence $\cot\theta = 1.7208$ or $-0.3874$.
$\theta = \operatorname{arccot} 1.7208 = 30°\ 10'$ or $210°\ 10'$, since cotangent is positive in the first and third quadrants,
or $\theta = \operatorname{arccot}(-0.3874) = 111°\ 11'$ or $291°\ 11'$, since cotangent is negative in the second and fourth quadrants.
Hence $\theta = \mathbf{30°\ 10', 111°\ 11', 210°\ 10'}$ or $\mathbf{291°\ 11'}$.

---

*Problem 20* Solve $7\sin^2\theta - 4\cos\theta = 5$ for values of $\theta$ between $0°$ and $360°$.

Since $\cos^2\theta + \sin^2\theta = 1$, $\sin^2\theta = 1 - \cos^2\theta$.
Substituting for $\sin^2\theta$ in $7\sin^2\theta - 4\cos\theta = 5$ gives $7(1 - \cos^2\theta) - 4\cos\theta = 5$
i.e. $7 - 7\cos^2\theta - 4\cos\theta = 5$
$\qquad -7\cos^2\theta - 4\cos\theta + 2 = 0$
$\qquad 7\cos^2\theta + 4\cos\theta - 2 = 0$.

Since the left hand side does not factorise the quadratic formula is used.

Thus $\cos\theta = \dfrac{-4 \pm \sqrt{\{(4)^2 - 4(7)(-2)\}}}{2(7)} = \dfrac{-4 \pm \sqrt{(16+56)}}{14} = \dfrac{-4 \pm \sqrt{72}}{14}$

$= \dfrac{4.4853}{14}$ or $\dfrac{-12.4853}{14} = 0.3204$ or $-0.8918$.

$\qquad \theta = \arccos 0.3204 = 71°\ 19'$ or $228°\ 41'$
or $\qquad \theta = \arccos(-0.8918) = 153°\ 6'$ or $206°\ 54'$.
Hence $\qquad \theta = \mathbf{71°\ 19', 153°\ 6', 206°\ 54'}$ or $\mathbf{288°\ 41'}$.

## C  FURTHER PROBLEMS ON TRIGONOMETRIC IDENTITIES AND THE SOLUTION OF TRIGONOMETRIC EQUATIONS

In *Problems 1 to 6* prove the trigonometric identities.

1. $\sin x \cot x = \cos x$.

2. $\dfrac{1}{\sqrt{(1-\cos^2 \theta)}} = \operatorname{cosec} \theta$

3. $2\cos^2 A - 1 = \cos^2 A - \sin^2 A$

4. $\dfrac{\cos x - \cos^3 x}{\sin x} = \sin x \cos x$

5. $(1+\cot \theta)^2 + (1-\cot \theta)^2 = 2\operatorname{cosec}^2 \theta$.

6. $\dfrac{\sin^2 x\,(\sec x + \operatorname{cosec} x)}{\cos x \tan x} = 1 + \tan x$

7. Show that the trigonometric identities $\cos^2 \theta + \sin^2 \theta = 1$, $1 + \tan^2 \theta = \sec^2 \theta$ and $\cot^2 \theta + 1 = \operatorname{cosec}^2 \theta$ are valid when $\theta$ is (a) $124°$, (b) $231°$ and (c) $312° \ 46'$.

In *Problems 8 to 17* solve the equations for angles between $0°$ and $360°$.

8. (a) $4 - 7\sin \theta = 0$     [(a) $\theta = 34° \ 51'$ or $145° \ 9'$
   (b) $2.5 \cos x + 1.75 = 0$     (b) $x = 134° \ 26'$ or $225° \ 34'$]

9. (a) $3 \operatorname{cosec} A + 5.5 = 0$     [(a) $A = 213° \ 3'$ or $326° \ 57'$
   (b) $4(2.32 - 5.4 \cot t) = 0$.     (b) $t = 66° \ 45'$ or $246° \ 45'$]

10. (a) $5 \sin^2 y = 3$;     [(a) $y = 50° \ 46'$, $129° \ 14'$, $230° \ 46'$ or $309° \ 14'$
    (b) $3 \tan^2 \phi - 2 = 0$     (b) $\phi = 39° \ 14'$, $140° \ 46'$, $219° \ 14'$ or $320° \ 46'$]

11. (a) $5 + 3 \operatorname{cosec}^2 D = 8$     [(a) $D = 90°$ or $270°$;
    (b) $2 \cot^2 \theta = 5$     (b) $\theta = 32° \ 19'$, $147° \ 41'$, $212° \ 19'$ or $327° \ 41'$]

12. (a) $15 \sin^2 A + \sin A - 2 = 0$     [(a) $A = 19° \ 28'$, $160° \ 32'$, $203° \ 35'$ or $336° \ 25'$
    (b) $8 \tan^2 \theta + 2 \tan \theta = 15$     (b) $\theta = 51° \ 20'$, $123° \ 41'$, $231° \ 20'$ or $303° \ 41'$]

13. (a) $\sec x + 6 = \sec^2 x$     [(a) $x = 70° \ 32'$, $120°$, $240°$ or $289° \ 28'$
    (b) $2 \operatorname{cosec}^2 t - 5 \operatorname{cosec} t = 12$     (b) $t = 14° \ 29'$, $165° \ 31'$, $221° \ 49'$ or $318° \ 11'$]

14. (a) $12 \sin^2 \theta - 6 = \cos \theta$     [(a) $\theta = 48° \ 11'$, $138° \ 35'$, $221° \ 25'$ or $311° \ 49'$
    (b) $16 \sec x - 2 = 14 \tan^2 x$     (b) $x = 52° \ 56'$ or $307° \ 4'$]

15. (a) $4 \cot^2 A - 6 \operatorname{cosec} A + 6 = 0$     [(a) $90°$
    (b) $2 \cos^2 y + 3 \sin y = 3$     (b) $30°$, $90°$ or $150°$]

16. (a) $3.2 \sin^2 x + 2.5 \cos x - 1.8 = 0$     [(a) $112° \ 11'$ or $247° \ 49'$
    (b) $5 \sec t + 2 \tan^2 t = 3$     (b) $107° \ 50'$ or $252° \ 10'$]

17. (a) $2.9 \cos^2 \alpha - 7 \sin \alpha + 1 = 0$     [(a) $\alpha = 27° \ 50'$ or $152° \ 10'$
    (b) $3 \operatorname{cosec}^2 \beta = 8 - 7 \cot \beta$     (b) $\beta = 60° \ 10'$, $161° \ 1'$, $240° \ 10'$ or $341° \ 1'$]

# 5 Areas and volumes

## A. MAIN POINTS CONCERNING AREAS AND VOLUMES

TABLE 1

**(i) Cylinder**

Volume $= \pi r^2 h$

Total surface area $= 2\pi rh + \pi r^2$

**(iii) Cone**

Volume $\frac{1}{3}\pi r^2 h$

Curved surface area $= \pi rl$
Total surface area $= \pi rl + \pi r^2$

**(ii) Pyramid**

Volume $= \frac{1}{3}Ah$ where $A =$ area of base

Total surface area = Sum of areas of triangles forming sides + Area of base

**(iv) Sphere**

Volume $= \frac{4}{3}\pi r^3$

Surface area $= 4\pi r^2$

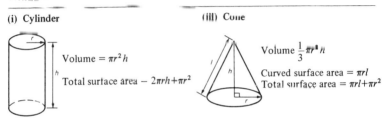
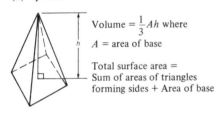
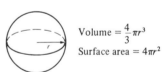

1 Volumes and surface areas of regular solids

2 (i) The **frustum** of a pyramid or cone is the portion remaining when a part containing the vertex is cut off by a plane parallel to the base.
 (ii) The **volume of a frustum of a pyramid or cone** is given by the volume of the whole pyramid or cone minus the volume of the small pyramid or cone cut off.
 (iii) The **surface area of the sides of a frustum of a pyramid or cone** is given by the surface area of the whole pyramid or cone minus the surface area of the small pyramid or cone cut off. This gives the lateral surface area of the frustum. If the total surface area of the frustum is required then the surface area of the two parallel ends are added to the lateral surface area.

(iv) There is an alternative method for finding the volume and surface area of a **frustum of a cone**. With reference to *Fig 1*:

Volume $= \frac{1}{3}\pi h (R^2+Rr+r^2)$

Curved surface area $= \pi l (R+r)$

Total surface area $= \pi l (R+r)+\pi r^2 +\pi R^2$

**Fig 1**

## B. WORKED PROBLEMS ON AREAS AND VOLUMES

*Problem 1* Determine the volume and total surface area of a square pyramid of perpendicular height 10.0 cm and length of side of base 6.20 cm.

The square pyramid is shown in *Fig 2*.
From *Table 1*, volume of pyramid,

$V = \frac{1}{3}$(area of base)(perpendicular height)

$= \frac{1}{3}(6.20)^2(10.0) =$ **128.1 cm³**

The total surface area consists of a square base and four equal triangles.

Area of triangle AQB in *Fig 2* is $\frac{1}{2}$(AB)(PQ).

Length PQ is calculated using Pythagoras' theorem on triangle OPQ.
$PQ^2 = OP^2 +OQ^2$, from which
$PQ =\sqrt{\{(3.10)^2 +(10.0)^2\}} = 10.47$ cm.

Area of triangle AQB $= \frac{1}{2}(6.20)(10.47)$
$= 32.46$ cm².

**Fig 2**

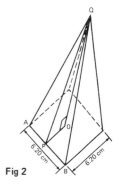

Total surface area of pyramid $= (6.20)^2 +4(32.46)$
$= 38.44+129.84 =$ **168.3 cm²**
(*Note*: Answers are given to 1 significant figure more than the least accurate number in the data.)

*Problem 2* Determine the volume and the total surface area of a cone of radius 4.50 cm and perpendicular height 12.5 cm.

From *Table 1*, Volume of cone $= \frac{1}{3}\pi r^2 h = \frac{1}{3}\pi(4.50)^2(12.5) =$ **265.1 cm³**

Total surface area of cone $= \pi r l+\pi r^2$, where $l$ is the slant height (see *Table 1*)
Using Pythagoras' theorem, slant height, $l =\sqrt{(h^2+r^2)} =\sqrt{\{(12.5)^2+(4.5)^2\}}$
$= 13.29$ cm

Hence total surface area $= \pi(4.50)(13.29)+\pi(4.50)^2$
$= 187.88+63.62 =$ **251.5 cm²**

*Problem 3* Find the volume and surface area of a sphere of diameter 7.20 cm.

From *Table 1*, Volume of sphere = $\frac{4}{3}\pi r^3 = \frac{4}{3}\pi \left(\frac{7.20}{2}\right)^3$ = **195.4 cm³**

Surface area of sphere = $4\pi r^2 = 4\pi \left(\frac{7.20}{2}\right)^2$ = **162.9 cm²**

*Problem 4* A pyramid has a rectangular base 3.60 cm by 5.40 cm. Determine the volume and total surface area of the pyramid if each of its sloping edges is 15.0 cm.

The pyramid is shown in *Fig 3*. To calculate the volume of the pyramid the perpendicular height EF is required. Diagonal BD is calculated using Pythagoras' theorem:

i.e.  BD = $\sqrt{\{(3.60)^2+(5.40)^2\}}$ = 6.490 cm

Hence EB = $\frac{1}{2}$BD = $\frac{6.490}{2}$ = 3.245 cm

Using Pythagoras' theorem on triangle BEF gives $BF^2 = EB^2 + EF^2$, from which, EF = $\sqrt{(BF^2 - EB^2)} = \sqrt{\{(15.0)^2 - (3.245)^2\}}$ = 14.64 cm.

Volume of pyramid = $\frac{1}{3}$(area of base)(perpendicular height)

$= \frac{1}{3}(3.60 \times 5.40)(14.64)$ = **94.87 cm³**

Area of triangle ADF (which equals triangle BCF) = $\frac{1}{2}$(AD)(FG), where G is the midpoint of AD.

Using Pythagoras' theorem on triangle FGA gives FG = $\sqrt{\{(15.0)^2 - (1.80)^2\}}$
$= 14.89$ cm

Hence area of triangle ADF = $\frac{1}{2}(3.60)(14.89)$ = 26.80 cm²

Similarly, if H is the mid-point of AB, then FH = $\sqrt{\{(15.0)^2 - (2.70)^2\}}$ = 14.75 cm

Hence area of triangle ABF (which equals triangle CDF) = $\frac{1}{2}(5.40)(14.75)$
$= 39.83$ cm²

Total surface area of pyramid = 2(26.80) + 2(39.83) + (3.60)(5.40)
$= 53.60 \quad + \quad 79.66 + \quad\quad 19.44 \quad =$ **152.7 cm²**

*Problem 5* Calculate the volume and total surface area of a hemisphere of diameter 5.0 cm.

Volume of hemisphere = $\frac{1}{2}$(volume of sphere) = $\frac{2}{3}\pi r^3 = \frac{2}{3}\pi\left(\frac{5.0}{2}\right)^3$ = **32.7 cm³**

Total surface area = curved surface area + area of circle

$= \frac{1}{2}$(surface area of sphere) + $\pi r^2$

$= 2\pi r^2 + \pi r^2 = 3\pi r^2 = 3\pi \left(\frac{5.0}{2}\right)^2$ = **58.9 cm²**

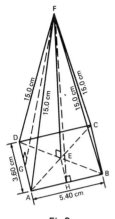

Fig 3

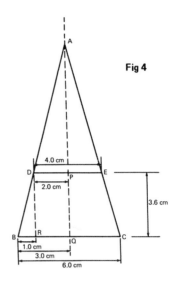

Fig 4

*Problem 6* Determine the volume of a frustum of a cone if the diameter of the ends are 6.0 cm and 4.0 cm and its perpendicular height is 3.6 cm.

*Method 1*
A section through the vertex of a complete cone is shown in *Fig 4*.

Using similar triangles $\frac{AP}{DP} = \frac{DR}{BR}$

Hence $\frac{AP}{2.0} = \frac{3.6}{1.0}$, from which AP = $\frac{(2.0)(3.6)}{1.0}$ = 7.2 cm

The height of the large cone = 3.6+7.2 = 10.8 cm
Volume of frustum of cone = volume of large cone−volume of small cone cut off.

$$= \frac{1}{3}\pi(3.0)^2(10.8) - \frac{1}{3}\pi(2.0)^2(7.2)$$

$$= 101.79 - 30.16 = \mathbf{71.6 \ cm^3}$$

*Method 2*
From para. 2(iv), volume of the frustum of a cone = $\frac{1}{3}\pi h\ (R^2 + Rr + r^2)$,
where $R$ = 3.0 cm, $r$ = 2.0 cm and $h$ = 3.6 cm

Hence volume of frustum = $\frac{1}{3}\pi(3.6)[(3.0)^2 + (3.0)(2.0) + (2.0)^2]$

$$= \frac{1}{3}\pi(3.6)(19.0) = \mathbf{71.6 \ cm^3}$$

*Problem 7* Find the total surface area of the frustum of the cone in *Problem 6*.

*Method 1*
Curved surface area of frustum = curved surface area of large cone − curved surface area of small cone cut off.
From *Fig 4*, using Pythagoras' theorem:

$AB^2 = AQ^2 + BQ^2$, from which $AB = \sqrt{\{(10.8)^2 + (3.0)^2\}} = 11.21$ cm
and $AD^2 = AP^2 + DP^2$, from which $AD = \sqrt{\{(7.2)^2 + (2.0)^2\}} = 7.47$ cm

Curved surface area of large cone = $\pi rl = \pi(BQ)(AB) = \pi(3.0)(11.21) = 105.7$ cm$^2$
and curved surface area of small cone = $\pi(DP)(AD) = \pi(2.0)(7.47) = 46.94$ cm$^2$
Hence curved surface area of frustum = $105.7 - 46.94 = 58.76$ cm$^2$
Total surface area of frustum = curved surface area + area of two circular ends
$= 58.76 + \pi(2.0)^2 + \pi(3.0)^2$
$= 58.76 + 12.57 + 28.27 = $ **99.6 cm$^2$**

*Method 2*
From para 2(iv), total surface area of frustum = $\pi l(R+r) + \pi r^2 + \pi R^2$
where $l = BD = 11.21 - 7.47 = 3.74$ cm, $R = 3.0$ cm and $r = 2.0$ cm
Hence total surface area of frustum = $\pi(3.74)(3.0 + 2.0) + \pi(2.0)^2 + \pi(3.0)^2$
$= $ **99.6 cm$^2$**

*Problem 8* A storage hopper is in the shape of a frustum of a pyramid. Determine its volume if the ends of the frustum are squares of sides 8.0 m and 4.6 m respectively and the perpendicular height between its ends is 3.6 m.

The frustum is shown shaded in *Fig 5(b)* as part of a complete pyramid. A section perpendicular to the base through the vertex is shown in *Fig 5(a)*.

By similar triangles: $\dfrac{CG}{BG} = \dfrac{BH}{AH}$

Hence $CG = BG \left(\dfrac{BH}{AH}\right) = \dfrac{(2.3)(3.6)}{(1.7)} = 4.87$ m

Height of complete pyramid = $3.6 + 4.87 = 8.47$ m

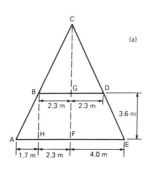
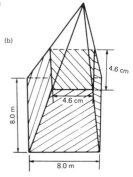

Fig 5

Volume of large pyramid $= \frac{1}{3}(8.0)^2(8.47) = 180.7 \text{ m}^3$

Volume of small pyramid cut off $= \frac{1}{3}(4.6)^2(4.87) = 34.35 \text{ m}^3$

Hence volume of storage hopper $= 180.7 - 34.35 = \mathbf{146.4 \text{ m}^3}$

*Problem 9* Determine the lateral surface area of the storage hopper in *Problem 8*.

The lateral surface area of the storage hopper consists of four equal trapeziums.

From *Fig 6*, area of trapezium PRSU

$= \frac{1}{2}(\text{PR}+\text{SU})(\text{QT})$

OT = 1.7 m (same as AH in *Fig 5(b)*) and
OQ = 3.6 m.
By Pythagoras' theorem,
QT $= \sqrt{(\text{OQ}^2+\text{OT}^2)} = \sqrt{\{(3.6)^2+(1.7)^2\}}$
$= 3.98$ m.

Area of trapezium PRSU

$= \frac{1}{2}(4.6+8.0)(3.98) = 25.07 \text{ m}^2$.

Lateral surface area of hopper $= 4(25.07)$
$= \mathbf{100.3 \text{ m}^2}$

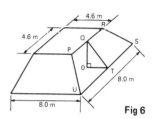

Fig 6

*Problem 10* A lampshade is in the shape of a frustum of a cone. The vertical height of the shade is 25.0 cm and the diameters of the ends are 20.0 cm and 10.0 cm respectively. Determine the area of the material needed to form the lampshade, correct to 3 significant figures.

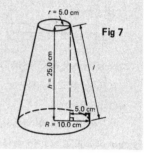

Fig 7

The curved surface area of a frustum of a cone $= \pi l(R+r)$, from para 2(iv). Since the diameters of the ends of the frustum are 20.0 cm and 10.0 cm, then, from *Fig 7*, $r = 5.0$ cm, $R = 10.0$ cm and $l = \sqrt{\{(25.0)^2+(5.0)^2\}} = 25.50$ cm, by Pythagoras' theorem. Hence curved surface area $= \pi(25.50)(10.0+5.0) = 1201.7 \text{ cm}^2$, i.e. the area of material needed to form the lampshade is $\mathbf{1200 \text{ cm}^2}$, correct to 3 significant figures.

*Problem 11* *Fig 8* shows a plan of a floor of a building which is to be carpeted. Calculate the area of the floor in square metres. Calculate the cost, correct to the nearest pound, of carpeting the floor with carpet costing £8.40 per m², assuming 30% of carpet is wasted in fitting.

Area of floor plan = area of triangle ABC + area of semicircle
+ area of rectangle CGLM + area of rectangle CDEF
− area of trapezium HIJK.

Triangle ABC is equilateral since AB = BC = 3 m and hence angle B'CB = 60°.

$\sin B'CB = \dfrac{BB'}{3}$, i.e. $BB' = 3 \sin 60° = 2.598$ m

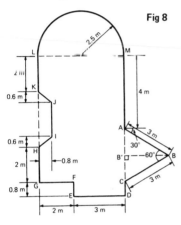

**Fig 8**

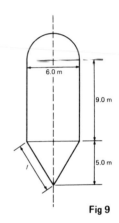

**Fig 9**

Area of triangle ABC = $\dfrac{1}{2}(AC)(BB')$
= $\dfrac{1}{2}(3)(2.598) = 3.897$ m²

Area of semicircle
= $\dfrac{1}{2}\pi r^2 = \dfrac{1}{2}\pi(2.5)^2 = 9.817$ m²

Area of CGLM = 5 × 7 = 35 m²
Area of CDEF = 0.8 × 3 = 2.4 m²

Area of HIJK = $\dfrac{1}{2}(KH+IJ)(0.8)$

Since MC = 7 m then LG = 7 m.
Hence JI = 7−5.2 = 1.8 m

Hence area of HIJK = $\dfrac{1}{2}(3+1.8)(0.8)$
= 1.92 m²

Total floor area = 3.897+9.817+35+2.4−1.92 = 49.194 m²
To allow for 30% wastage, amount of carpet required = 1.3 × 49.194 = 63.95 m²
Cost of carpet at £8.40 per m² = 63.95 × 8.40 = **£537**, correct to the nearest pound

*Problem 12* A boiler consists of a cylindrical section of length 9.0 m and diameter 6.0 m, on one end of which is surmounted a hemispherical section of diameter 6.0 m and on the other end a conical section of height 5.0 m. Determine the volume of the boiler.

A section through the boiler is shown in *Fig 9*.
The radius of the hemisphere, cylinder and cone is $6.0/2 = 3.0$ m.

Volume of hemisphere $= \frac{2}{3}\pi r^3 = \frac{2}{3}\pi(3.0)^3 = 56.55$ m$^3$

Volume of cylinder $= \pi r^2 h = \pi(3.0)^2(9.0) = 254.5$ m$^3$

Volume of cone $= \frac{1}{3}\pi r^2 h = \frac{1}{3}\pi(3.0)^2(5.0) = 47.12$ m$^3$

**Total volume of boiler** $= 56.55 + 254.5 + 47.12 =$ **358 m$^3$**

*Problem 13* Find the total surface area of the boiler in *Problem 12*.

Curved surface area of hemisphere $= 2\pi r^2 = 2\pi(3.0)^2 = 56.55$ m$^2$
Curved surface area of cylinder $= 2\pi rh = 2\pi(3.0)(9.0) = 169.65$ m$^2$
Curved surface area of cone $= \pi rl = \pi(3.0)[\sqrt{\{(5.0)^2 + (3.0)^2\}}] = 54.96$ m$^2$
Total surface area of boiler $= 56.55 + 169.65 + 54.96 =$ **281 m$^2$**

*Problem 14* A cooling tower is in the form of a cylinder surmounted by a frustum of a cone as shown in *Fig 10*. Determine the volume of air space in the tower if 40% of the space is used for pipes and other structures.

Volume of cylindrical portion $= \pi r^2 h = \pi(25.0/2)^2(12.0) = 5890$ m$^3$

Volume of frustum of cone $= \frac{1}{3}\pi h(R^2 + Rr + r^2)$,

where $h = 30.0 - 12.0 = 18.0$ m, $R = 25.0/2 = 12.5$ m and $r = 12.0/2 = 6.0$ m.

Hence volume of frustum of cone

$= \frac{1}{3}\pi(18.0)[(12.5)^2 + (12.5)(6.0) + (6.0)^2]$
$= 5038$ m$^3$.

Total volume of cooling tower
$= 5890 + 5038 = 10\ 928$ m$^3$.

If 40% of space is occupied then volume of
air space $= 0.6 \times 10\ 928 =$ **6557 m$^3$**

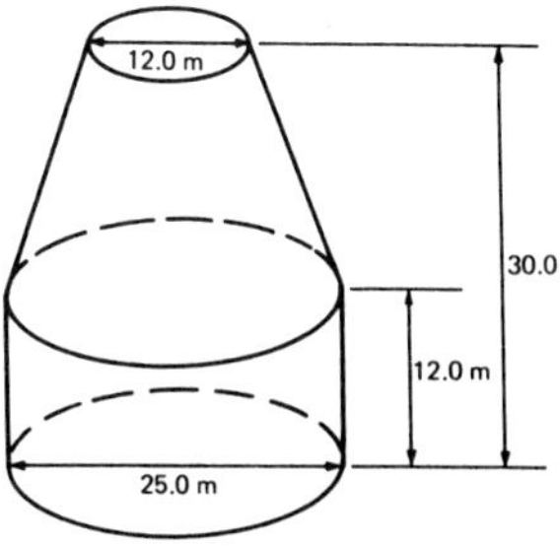

**Fig 10**

## C. FURTHER PROBLEMS ON AREAS AND VOLUMES

1. Determine (a) the volume and (b) the total surface area of the following solids:
   (i) a cone of radius 8.0 cm and perpendicular height 10 cm.
   $$[(a)\ 670\ cm^3\ ;(b)\ 523\ cm^2]$$
   (ii) a sphere of diameter 7.0 cm. $[(a)\ 180\ cm^3\ ;(b)\ 154\ cm^2]$
   (iii) a hemisphere of radius 3.0 cm. $[(a)\ 56.5\ cm^3\ ;(b)\ 84.8\ cm^2]$
   (iv) a 2.5 cm by 2.5 cm square pyramid of perpendicular height 5.0 cm.
   $$[(a)\ 10.4\ cm^3\ ;(b)\ 32.0\ cm^2]$$
   (v) a 4.0 cm by 6.0 cm rectangular pyramid of perpendicular height 12.0 cm.
   $$[(a)\ 96.0\ cm^3\ ;(b)\ 146\ cm^2]$$
   (vi) a 4.2 cm by 4.2 cm square pyramid whose sloping edges are each 15.0 cm.
   $$[(a)\ 86.5\ cm^3\ ;(b)\ 142\ cm^2]$$
   (vii) a pyramid having an octagonal base of side 5.0 cm and perpendicular height 20 cm.
   $$[(a)\ 805\ cm^3\ ;(b)\ 539\ cm^2]$$

2. The volume of a sphere is 325 cm³. Determine its diameter.
   $$[8.53\ cm]$$

3. A metal sphere weighing 24 kg is melted down and recast into a solid cone of base radius 8.0 cm. If the density of the metal is 8000 kg/m³, determine (a) the diameter of the metal sphere and (b) the perpendicular height of the cone, assuming that 15% of the metal is lost in the process.
   $$[(a)\ 17.9\ cm;(b)\ 38.0\ cm]$$

4. Find the volume of a regular hexagonal pyramid if the perpendicular height is 16.0 cm and the side of base is 3.0 cm.
   $$[125\ cm^3]$$

5. The radii of the faces of a frustum of a cone are 2.0 cm and 4.0 cm and the thickness of the frustum is 5.0 cm. Determine its volume and total surface area.
   $$[147\ cm^3\ ;\ 164\ cm^2]$$

6. A frustum of a pyramid has square ends, the squares having sides 9.0 cm and 5.0 cm respectively. Calculate the volume and total surface area of the frustum if the perpendicular distance between its ends is 8.0 cm.
   $$[403\ cm^3\ ;\ 337\ cm^2]$$

7. A cooling tower is in the form of a frustum of a cone. The base has a diameter of 32.0 m, the top has a diameter of 14.0 m and the vertical height is 24.0 m. Calculate the volume of the tower and the curved surface area.
   $$[10\ 480\ m^3\ ;\ 1852\ m^2]$$

8. A loudspeaker diaphragm is in the form of a frustum of a cone. If the end diameters are 28.0 cm and 6.00 cm and the vertical distance between the ends is 30.0 cm, find the area of material needed to cover the curved surface of the speaker.
   $$[1707\ cm^2]$$

9. A rectangular prism of metal having dimensions 4.3 cm by 7.2 cm by 12.4 cm is melted down and recast into a frustum of a square pyramid, 10% of the metal being lost in the process. If the ends of the frustum are squares of size 3 cm and 8 cm respectively, find the thickness of the frustum.
   $$[10.69\ cm]$$

10. Determine the volume and total surface area of a bucket consisting of an inverted frustum of a cone, of slant height 36.0 cm and end diameters 55.0 cm and 35.0 cm.
    $$[55\ 910\ cm^3\ ;\ 6051\ cm^2]$$

11. A cylindrical tank of diameter 2.0 m and perpendicular height 3.0 m is to be replaced by a tank of the same capacity but in the form of a frustum of a cone. If

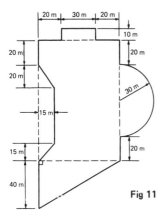

Fig 11

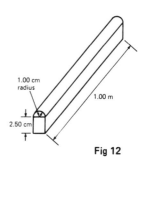

Fig 12

the diameters of the ends of the frustum are 1.0 m and 2.0 m respectively, determine the vertical height required.

[5.14 m]

12 A plot of land is in the shape shown in *Fig 11*. Determine (a) its area in hectares (1 ha = $10^4$ m$^2$), and (b) the length of fencing required, to the nearest metre, to completely enclose the plot of land.

[(a) 0.918 ha; (b) 456 m]

13 A buoy consists of a hemisphere surmounted by a cone. The diameter of the cone and hemisphere is 2.5 m and the slant height of the cone is 4.0 m. Determine the volume and surface area of the buoy.

[10.3 m$^3$; 25.5 m$^2$]

14 A petrol container is in the form of a central cylindrical portion 5.0 m long with a hemispherical section surmounted on each end. If the diameters of the hemisphere and cylinder are both 1.2 m determine the capacity of the tank in litres.
(1 litre = 1000 cm$^3$) [6560 l]

15 A 12.0 m high marquee is in the shape of a cylinder surmounted by a cone. The cylindrical portion has a height of 5.0 m with a diameter of 20.0 m. Determine (a) the amount of canvas needed to make the marquee assuming a 10% wastage and (b) its cost, correct to the nearest pound, if canvas costs £2.60 per m$^2$.

[(a) 767.4 m$^2$; (b) £1995]

16 *Fig 12* shows a metal rod section. Determine its volume and total surface area.
[657.1 cm$^3$; 1027 cm$^2$]

# 6 Irregular areas and volumes and mean values of waveforms

## A. MAIN POINTS CONCERNED WITH IRREGULAR AREAS AND VOLUMES AND MEAN VALUES OF WAVEFORMS

1. **Areas of irregular figures**

   Areas of irregular plane surfaces may be approximately determined by using (a) a planimeter, (b) a trapezoidal rule, (c) the mid-ordinate rule, and (d) Simpson's rule. Such methods may be used, for example, by engineers estimating areas of indicator diagrams of steam engines, surveyors estimating areas of plots of land or naval architects estimating areas of water planes or transverse sections of ships.

   (a) **A planimeter** is an instrument for directly measuring small areas bounded by an irregular curve.

   (b) **Trapezoidal rule.**

   To determine the area PQRS in *Fig 1*:
   
   (i) Divide base PS into any number of equal intervals, each of width $d$, (the greater the number of intervals, the greater the accuracy).
   
   (ii) Accurately measure ordinates $y_1, y_2, y_3$, etc.
   
   (iii) Area PQRS = $d \left[ \dfrac{y_1 + y_7}{2} + y_2 + y_3 + y_4 + y_5 + y_6 \right]$

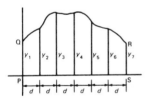

**Fig 1**

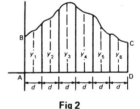

**Fig 2**

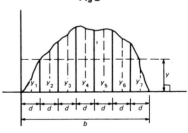

**Fig 3 (above)**
**Fig 4 (right)**

In general, the trapezoidal rule states:

**Area = (width of interval)$\left[\frac{1}{2}(\text{first + last ordinate}) + \text{sum of remaining ordinates}\right]$**

(c) **Mid-ordinate rule**

To determine the area ABCD of *Fig 2*:
(i) Divide base AD into any number of equal intervals, each of width $d$, (the greater the number of intervals, the greater the accuracy).
(ii) Erect ordinates in the middle of each interval (shown by broken lines in *Fig 2*).
(iii) Accurately measure ordinates $y_1, y_2, y_3$, etc.
(iv) Area ABCD = $d(y_1+y_2+y_3+y_4+y_5+y_6+y_7)$

In general, the mid-ordinate rule states:

**Area = (width of interval)(sum of mid-ordinates)**

(d) **Simpson's rule**

To determine the area PQRS of *Fig 1*:
(i) Divide base PS into an even number of intervals, each of width $d$, (the greater the number of intervals, the greater the accuracy).
(ii) Accurately measure ordinates $y_1, y_2, y_3$, etc.
(iii) Area PQRS = $\frac{d}{3}\left[(y_1+y_7) + 4(y_2+y_4+y_6) + 2(y_3+y_5)\right]$

In general, Simpson's rule states:

$$\text{Area} = \frac{1}{3}(\text{width of interval})\left[\binom{\text{first + last}}{\text{ordinate}} + 4\binom{\text{sum of even}}{\text{ordinates}} + 2\binom{\text{sum of remaining}}{\text{odd ordinates}}\right]$$

2 **Volume of irregular solids**

If the cross-sectional areas $A_1, A_2, A_3$, etc., of an irregular solid bounded by two parallel planes are known at equal intervals of width $d$ (as shown in *Fig 3*), then by Simpson's rule:

**Volume, $V = \frac{d}{3}[(A_1+A_7) + 4(A_2+A_4+A_6) + 2(A_3+A_5)]$**

3 **Mean or average value of a waveform.**

(i) The mean or average value, $y$, of the waveform shown in *Fig 4* is given by:

$$y = \frac{\text{area under curve}}{\text{length of base, } b}$$

(ii) If the mid-ordinate rule is used to find the area under the curve, then:

$$y = \frac{\text{sum of mid-ordinates}}{\text{number of mid-ordinates}}$$

(a)

(b)

$\left(=\dfrac{y_1+y_2+y_3+y_4+y_5+y_6+y_7}{7}\quad \text{for } Fig\ 4\right)$

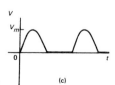

(c)

**Fig 5**

(iii) For a **sine-wave**, the mean or average value:
   (i) over one complete cycle is zero (see *Fig 5(a)*),
   (ii) over half a cycle is 0.637 × maximum value, or $2/\pi$ × maximum value,
   (iii) of a full-wave rectified waveform (see *Fig 5(b)*) is 0.637 × maximum value,
   (iv) of a half-wave rectified waveform (see *Fig 5(c)*) is 0.318 × maximum value, or $1/\pi$ × maximum value.

## B. WORKED PROBLEMS ON IRREGULAR AREAS AND VOLUMES AND MEAN VALUES OF WAVEFORMS

*Problem 1* A car starts from rest and its speed is measured every second for 6 s.

| Time $t$ (s)   | 0 | 1   | 2   | 3    | 4    | 5    | 6    |
|----------------|---|-----|-----|------|------|------|------|
| Speed $v$ (m/s)| 0 | 2.5 | 5.5 | 8.75 | 12.5 | 17.5 | 24.0 |

Determine the distance travelled in 6 seconds (i.e. the area under the $v/t$ graph) by (a) the trapezoidal rule, (b) the mid-ordinate rule, and (c) Simpson's rule.

A graph of speed/time is shown in *Fig 6*.

(a) **Trapezoidal rule** (see para 1(b)).
The time base is divided into 6 strips each of width 1 s, and the length of the ordinates measured.

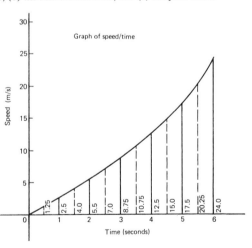

**Fig 6**

Thus area = $(1)\left[\left(\dfrac{0+24.0}{2}\right) + 2.5+5.5+8.75+12.5+17.5\right]$ = **58.75 m**

(b) **Mid-ordinate rule** (see para 1(c)).
The time base is divided into 6 strips each of width 1 second. Mid-ordinates are erected as shown in *Fig 6* by the broken lines. The length of each mid-ordinate is measured.

Thus area = (1)[1.25+4.0+7.0+10.75+15.0+20.25] = **58.25 m**

(c) **Simpson's rule** (see para 1(d)).
The time base is divided into 6 strips each of width 1 s, and the length of the ordinates measured.

Thus area = $\frac{1}{3}$(1)[(0+24.0)+4(2.5+8.75+17.5)+2(5.5+12.5)] = **58.33 m**

*Problem 2* A river is 15 m wide. Soundings of the depth are made at equal intervals of 3 m across the river and are as shown below.
Depth (m)   0   2.2   3.3   4.5   4.2   2.4   0
Calculate the cross-sectional area of the flow of water at this point using Simpson's rule.

From para 1(d), Area = $\frac{1}{3}$(3)[(0+0)+4(2.2+4.5+2.4)+2(3.3+4.2)]
= (1)[0+36.4+15] = **51.4 m²**

*Problem 3* Determine the average values over half a cycle of the periodic waveforms shown in *Fig 7*.

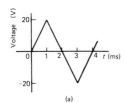
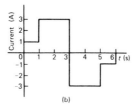
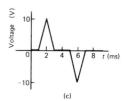

**Fig 7**

(a) Area under triangular waveform (a) for a half cycle is given by:

Area = $\frac{1}{2}$(base)(perpendicular height)

= $\frac{1}{2}$(2 × 10⁻³)(20) = 20 × 10⁻³ V s.

Average value of waveform = $\frac{\text{area under curve}}{\text{length of base}}$ = $\frac{20 \times 10^{-3} \text{ V s}}{2 \times 10^{-3} \text{ s}}$ = **10 V**

(b) Area under waveform (b) for a half cycle = (1 × 1)+(3 × 2) = 7 A s.

Average value of waveform = $\frac{\text{area under curve}}{\text{length of base}}$ = $\frac{7 \text{ A s}}{3 \text{ s}}$ = **2.33 A**

(c) A half cycle of the voltage waveform (c) is completed in 4 ms.

Area under curve = $\frac{1}{2}${(3−1)10⁻³}(10) = 10 × 10⁻³ V s

Average value of waveform = $\frac{\text{area under curve}}{\text{length of base}}$ = $\frac{10 \times 10^{-3} \text{ V s}}{4 \times 10^{-3} \text{ s}}$ = **2.5 V**

*Problem 4* Determine the mean value of current over one complete cycle of the periodic waveforms shown in *Fig 8*.

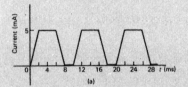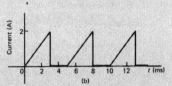

**Fig 8**

(a) One cycle of the trapezoidal waveform (a) is completed in 10 ms (i.e. the periodic time is 10 ms).

Area under curve = area of trapezium

$$= \frac{1}{2}(\text{sum of parallel sides})(\text{perpendicular distance between parallel sides})$$

$$= \frac{1}{2}\{(4+8) \times 10^{-3}\}(5 \times 10^{-3})$$

$$= 30 \times 10^{-6} \text{ A s}.$$

Mean value over one cycle = $\frac{\text{area under curve}}{\text{length of base}} = \frac{30 \times 10^{-6} \text{ A s}}{10 \times 10^{-3} \text{ s}} = $ **3 mA**

(b) One cycle of the sawtooth waveform (b) is completed in 5 ms.

Area under curve = $\frac{1}{2}(3 \times 10^{-3})(2) = 3 \times 10^{-3}$ A s

Mean value over one cycle = $\frac{\text{area under curve}}{\text{length of base}} = \frac{3 \times 10^{-3} \text{ A s}}{5 \times 10^{-3} \text{ s}} = $ **0.6 A**

---

*Problem 5* The power used in a manufacturing process during a 6 hour period is recorded at intervals of 1 hour as shown below.

| Time (h) | 0 | 1 | 2 | 3 | 4 | 5 | 6 |
|---|---|---|---|---|---|---|---|
| Power (kW) | 0 | 14 | 29 | 51 | 45 | 23 | 0 |

Plot a graph of power against time and, by using the mid-ordinate rule, determine (a) the area under the curve and (b) the average value of the power.

The graph of power/time is shown in *Fig 9*.

(a) The time base is divided into 6 equal intervals, each of width 1 hour. Mid-ordinates are erected (shown by broken lines in *Fig 9*) and measured. The values are shown in *Fig 9*.

Area under curve = (width of interval)(sum of mid-ordinates)
= (1)(7.0+21.5+42.0+49.5+37.0+10.0)
= **167 kW h** (i.e. a measure of electrical energy)

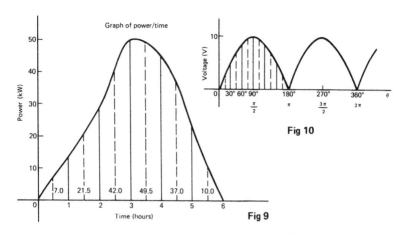

Fig 10

Fig 9

(b) Average value of waveform = $\dfrac{\text{area under curve}}{\text{length of base}} = \dfrac{167 \text{ kW h}}{6 \text{ h}} = 27.83$ kW

Alternatively, average value = $\dfrac{\text{sum of mid-ordinates}}{\text{number of mid-ordinates}}$

*Problem 6* Fig 10 shows a sinusoidal output voltage of a full-wave rectifier. Determine, using the mid-ordinate rule with 6 intervals, the mean output voltage.

One cycle of the output voltage is completed in $\pi$ radians or $180°$.
The base is divided into 6 intervals, each of width $30°$. The mid-ordinate of each interval will lie at $15°$, $45°$, $75°$, etc.
At $15°$ the height of the mid-ordinate is $10 \sin 15° = 2.588$ V.
At $45°$ the height of the mid-ordinate is $10 \sin 45° = 7.071$ V, and so on.
The results are tabulated below:

| Mid-ordinate | Height of mid-ordinate |
|---|---|
| $15°$ | $10 \sin 15° = 2.588$ V |
| $45°$ | $10 \sin 45° = 7.071$ V |
| $75°$ | $10 \sin 75° = 9.659$ V |
| $105°$ | $10 \sin 105° = 9.659$ V |
| $135°$ | $10 \sin 135° = 7.071$ V |
| $165°$ | $10 \sin 165° = 2.588$ V |

Sum of mid-ordinates = 38.636 V

Mean or average value of output voltage = $\dfrac{\text{sum of mid-ordinates}}{\text{number of mid-ordinates}} = \dfrac{38.636}{6}$

= **6.439 V**

(With a larger number of intervals a more accurate answer may be obtained. For a sine wave, the actual mean value is $0.637 \times$ maximum value, which in this problem gives 6.37 V.)

*Problem 7* An indicator diagram for a steam engine is shown in *Fig 11*. The base line has been divided into 6 equally spaced intervals and the lengths of the 7 ordinates measured with the results shown in centimetres. Determine (a) the area of the indicator diagram using Simpson's rule, and (b) the mean pressure in the cylinder given that 1 cm represents 100 kPa.

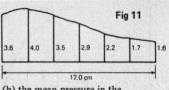

Fig 11

(a) The width of each interval is $12.0/6 = 2.0$ cm.

Using Simpson's rule, area = $\frac{1}{3}(2.0)[(3.6+1.6)+4(4.0+2.9+1.7)+2(3.5+2.2)]$

$= \frac{2}{3}[5.2+34.4+11.4] = \mathbf{34 \text{ cm}^2}$

(b) Mean height of ordinates = $\frac{\text{area of diagram}}{\text{length of base}} = \frac{34}{12} = 2.83$ cm

Since 1 cm represents 100 kPa, the mean pressure in the cylinder
= 2.83 cm × 100 kPa/cm

= **283 kPa**

*Problem 8* A tree trunk is 12 m in length and has a varying circular cross-section. The cross-sectional areas at intervals of 2 m measured from one end are: 0.52, 0.55, 0.59, 0.63, 0.72, 0.84, 0.97 m². Estimate the volume of the tree trunk.

A sketch of the tree trunk is similar to that shown in *Fig 3*, page 58, where $d = 2$ m, $A_1 = 0.52$ m², $A_2 = 0.55$ m², and so on. Using Simpson's rule for volumes (see para. 2) gives:

Volume = $\frac{2}{3}[(0.52+0.97)+4(0.55+0.63+0.84)+2(0.59+0.72)]$

$= \frac{2}{3}[1.49+8.08+2.62] = \mathbf{8.13 \text{ m}^3}$

*Problem 9* The areas of 7 horizontal cross-sections of a water reservoir at intervals of 10 m are: 210, 250, 320, 350, 290, 230, 170 m². Calculate the capacity of the reservoir in litres.

Using Simpson's rule for volumes (see para. 2) gives:
Volume = $\frac{10}{3}[(210+170)+4(250+350+230)+2(320+290)]$

$= \frac{10}{3}[380+3320+1220] = \mathbf{16\,400 \text{ m}^3}$.

$16\,400$ m³ $= 16\,400 \times 10^6$ cm³.

Since 1 litre = 1000 cm$^3$, capacity of reservoir = $\dfrac{16\,400 \times 10^6}{1000}$ litres

= 16 400 000 = **1.64 × 10$^7$ litres**

## C. FURTHER PROBLEMS ON IRREGULAR AREAS AND VOLUMES AND MEAN VALUES OF WAVEFORMS

1  Plot a graph of $y = 3x - x^2$ by compiling a table of values of $y$ from $x = 0$ to $x = 3$. Determine the area enclosed by the curve, the $x$-axis and ordinate $x = 0$ and $x = 3$ by (a) the trapezoidal rule, (b) the mid-ordinate rule and (c) by Simpson's rule.
   [4½ square units]

2  Plot the graph of $y = 2x^2 + 3$ between $x = 0$ and $x = 4$. Estimate the area enclosed by the curve, the ordinates $x = 0$ and $x = 4$, and the $x$-axis by an approximate method.
   [54.7 square units]

3  The velocity of a car at one second intervals are given in the following table:

| time $t$ (s)        | 0 | 1   | 2   | 3   | 4    | 5    | 6    |
|---------------------|---|-----|-----|-----|------|------|------|
| velocity $v$ (m/s)  | 0 | 2.0 | 4.5 | 8.0 | 14.0 | 21.0 | 29.0 |

Determine the distance travelled in 6 seconds (i e the area under the $v/t$ graph) using an approximate method.
   [63 m]

4  The shape of a piece of land is shown in *Fig 12*. To estimate the area of the land, a surveyor takes measurements at intervals of 50 m, perpendicular to the straight portion with the results shown (the dimensions being in metres). Estimate the area of the land in hectares (1 ha = $10^4$ m$^2$).
   [4.70 ha]

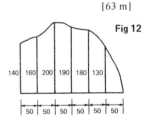

Fig 12

5  Determine the mean value of the periodic waveforms shown in *Fig 13* over a half cycle.
   [(a) 2 A; (b) 50 V; (c) 2.5 A]

6  The deck of a ship is 35 m long. At equal intervals of 5 m the width is given by the following table:

| Width (m) | 0 | 2.8 | 5.2 | 6.5 | 5.8 | 4.1 | 3.0 | 2.3 |

Estimate the area of the deck.
   [143 m$^2$]

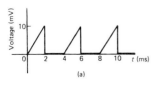

(a)

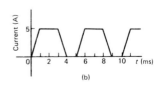

(b)

Fig 13

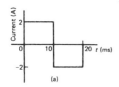
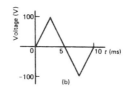
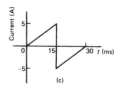

**Fig 14**

7  Find the average value of the periodic waveforms shown in *Fig 14* over one complete cycle.

[(a) 2.5 V; (b) 3 A]

8  An alternating current has the following values at equal intervals of 5 ms.

| Time (ms)    | 0 | 5   | 10  | 15  | 20  | 25  | 30 |
|---|---|---|---|---|---|---|---|
| Current (A)  | 0 | 0.9 | 2.6 | 4.9 | 5.8 | 3.5 | 0  |

Plot a graph of current against time and estimate the area under the curve over the 30 ms period using the mid-ordinate rule and determine its mean value.

[0.093 A s; 3.1 A]

9  Determine, using an approximate method, the average value of a sine wave of maximum value 50 V for (a) a half cycle and (b) a complete cycle.

[(a) 31.83 V; (b) 0]

10 An indicator diagram of a steam engine is 12 cm long. Seven evenly spaced ordinates, including the end ordinates, are measured as follows:

5.90, 5.52, 4.22, 3.63, 3.32, 3.24, 3.16 cm.

Determine the area of the diagram and the mean pressure in the cylinder if 1 cm represents 90 kPa.

[49.13 cm², 368.5 kPa]

11 The areas of equidistantly spaced sections of the underwater form of a small boat are as follows:

1.76, 2.78, 3.10, 3.12, 2.61, 1.24, 0.85 m²

Determine the underwater volume if the sections are 3 m apart.

[42.59 m³]

12 To estimate the amount of earth to be removed when constructing a cutting the cross-sectional area at intervals of 8 m were estimated as follows:

0, 2.8, 3.7, 4.5, 4.1, 2.6, 0 m³

Estimate the volume of earth to be excavated.

[147 m³]

13 The circumference of a 12 m long log of timber of varying circular cross-section is measured at intervals of 2 m along its length and the results are:

| Distance from one end (m) | 0    | 2    | 4    | 6    | 8    | 10   | 12   |
|---|---|---|---|---|---|---|---|
| Circumference (m)         | 2.80 | 3.25 | 3.94 | 4.32 | 5.16 | 5.82 | 6.36 |

Estimate the volume of the timber in cubic metres.

[20.42 m³]

# 7 Centroids of simple shapes

## A. MAIN POINTS CONCERNED WITH CENTROIDS OF SIMPLE SHAPES

1. A **lamina** is a thin, flat sheet having uniform thickness. The **centre of gravity** of a lamina is the point where it balances perfectly, i.e. the lamina's **centre of mass**. When dealing with a shape or area (i.e. a lamina of negligible thickness and mass) the term **centre of area** or **centroid** is used for the point where the centre of gravity of a lamina of that shape would lie.
2. (i) The centroid C of a **rectangle** lies on the intersection of the diagonals (see *Fig 1(a)*.
   (ii) The centroid C of a **triangle** lies on the intersection of its medians, a median being a line which joins the vertices of a triangle with the mid-point of the opposite side. It may be shown that the centroid lies at one-third of the perpendicular height above any side as base (see *Fig 1(b)*).
   (iii) The centroid C of a **circle** lies at its centre (see *Fig 1(c)*).
   (iv) The centroid C of a **semicircle** of radius $r$ lies on the centre line at a distance $4r/3\pi$ from the diameter (see *Fig 1(d)*).
3. The **first moment of area** is defined as the product of the area and the perpendicular distance of its centroid from a given axis in the plane of the area. In *Fig 2*, the first moment of area $A$ about axis XX is given by $(Ay)$ cubic units.
4. A **composite area** consists of two or more areas having different shapes joined together. The centroid of a composite area is found by dividing the whole area into parts, the centroids of which are known, and then taking moments (i.e. finding the first moment of area) about two orthogonal axes (i.e two axes lying

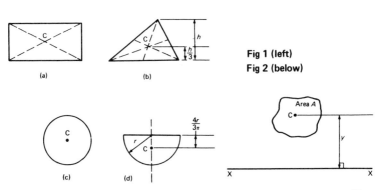

Fig 1 (left)
Fig 2 (below)

in the same place and at right angles to each other) For the composite area shown in *Fig 3*:

Sum of moments about YY, $\Sigma ax = a_1 x_1 + a_2 x_2 + a_3 x_3$
Sum of moments about XX, $\Sigma ay = a_1 y_1 + a_2 y_2 + a_3 y_3$
If $A = a_1 + a_2 + a_3$ and $\bar{x}$ and $\bar{y}$ are the distances of the centroid of the composite area about axes YY and XX respectively, then:

$$A\bar{x} = \Sigma ax, \text{ from which } \bar{x} = \frac{\Sigma ax}{A} = \frac{\text{first moment of area about YY}}{\text{total area}}$$

and $A\bar{y} = \Sigma ay$, from which $\bar{y} = \dfrac{\Sigma ay}{A} = \dfrac{\text{first moment of area about XX}}{\text{total area}}$

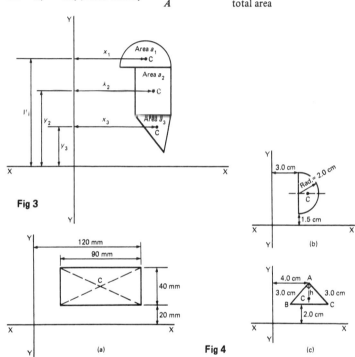

**Fig 3**

**Fig 4**

## B. WORKED PROBLEMS ON CENTROIDS OF SIMPLE SHAPES

*Problem 1* Determine the first moment of area of each of the shapes shown in *Fig 4* about axes XX and YY.

(a) Area of rectangle = 90 × 40 = 3600 mm²

Distance of centroid C of rectangle from XX = $20 + \dfrac{40}{2} = 40$ mm

First moment of area of rectangle about XX = area × perpendicular distance to centroid
$= (3600)(40) = \mathbf{144\ 000\ mm^3}$

Distance of centroid C of rectangle from YY $= (120-90) + \dfrac{90}{2} = 75$ mm

First moment of area of rectangle about YY $= (3600)(75) = \mathbf{270\ 000\ mm^3}$

(b) Area of semicircle $= \dfrac{\pi r^2}{2} = \dfrac{\pi (2)^2}{2} = 2\pi$ cm$^2$

Distance of centroid C of semicircle from XX $= 1.5 + 2.0 = 3.5$ cm.

First moment of area of semicircle about XX $= (2\pi)(3.5) = 7\pi = \mathbf{22.0\ cm^3}$

Distance of centroid C of semicircle from YY $= 3.0 + \dfrac{4r}{3\pi} = 3.0 + \dfrac{4(2.0)}{3\pi}$
$= 3.849$ cm

First moment of area of semicircle about YY $= (2\pi)(3.849) = \mathbf{24.2\ cm^3}$

(c) Area of triangle ABC $= \dfrac{1}{2}(3.0)(3.0) = 4.5$ cm$^2$.

Since the triangle ABC is isosceles $\angle B = \angle C = 45°$. Hence $\sin 45° = h/3.0$, from which $h = 3.0 \sin 45° = 2.121$ cm.

Distance of centroid C from XX $= 2.0 + \dfrac{h}{3.0} = 2.0 + \dfrac{2.121}{3.0} = 2.707$ cm

First moment of area of triangle about XX $= (4.5)(2.707) = \mathbf{12.2\ cm^3}$

Distance of centroid C from YY $= 4.0$ cm

First moment of area about YY $= (4.5)(4.0) = \mathbf{18.0\ cm^3}$

*Problem 2* Determine the position of the centroid of the symmetrical T-section shown in *Fig 5*.

Since the T-section is symmetrical the centroid will lie on axis YY. The T-section is divided into rectangles A and B with centroids at $C_A$ and $C_B$ respectively. Let the centroid, C, of the whole area be at a distance $\bar{y}$ from axis XX.

Area of rectangle A $= (4.0)(20.0) = 80.0$ cm$^2$
Area of rectangle B $= (5.0)(18.0) = 90.0$ cm$^2$
Total area A of T-section $= 80.0 + 90.0 = 170.0$ cm$^2$.

First moment of area of rectangle A about XX = (area) × (perpendicular distance to centroid of A from XX)

$$= (80.0)(5.0 + \dfrac{20.0}{2}) = 1200\ \text{cm}^3$$

First moment of area of rectangle B about XX $= (90.0)\left(\dfrac{5.0}{2}\right) = 225$ cm$^3$

Hence, from para 4, $A\bar{y} = \Sigma ax = 1200 + 225 = 1425$ cm$^3$

i.e. $\bar{y} = \dfrac{1425}{A} = \dfrac{1425}{170.0} = 8.38$ cm.

**Hence the centroid of the T-section lies on the axis of symmetry 8.38 cm from the base.**

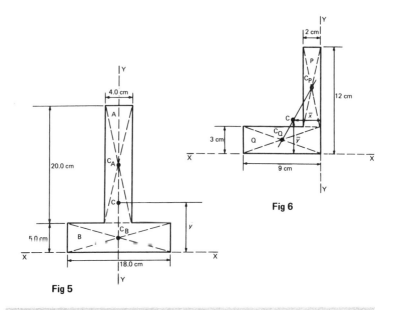

Fig 5

Fig 6

*Problem 3* Calculate the position of the centroid of the angle-iron section shown in *Fig 6* about axes XX and YY.

The L-section is divided into two rectangles P and Q with centroids at $C_P$ and $C_Q$ respectively. Let the centroid, C, of the whole area be at a perpendicular distance $\bar{x}$ from axis YY and at a perpendicular distance $\bar{y}$ from axis XX.

Area of rectangle P = $(12-3)(2)$ = 18 cm$^2$
Area of rectangle Q = $(9)(3)$ = 27 cm$^2$
Total area, $A$, of L-section = $18+27 = 45$ cm$^2$

The first moment of area of rectangle P about YY = (area)(perpendicular distance to centroid of P from YY)
$= (18)(1) = 18$ cm$^3$

The first moment of area of rectangle Q about YY = $(27)(4.5) = 121.5$ cm$^3$.
From para. 4, $A\bar{x} = \Sigma ax = 18 + 121.5 = 139.5$ cm$^3$.

Hence $\bar{x} = \dfrac{139.5}{A} = \dfrac{139.5}{45} = 3.1$ cm.

The first moment of area of rectangle P about XX = $(18)(3+4.5) = 135$ cm$^3$
The first moment of area of rectangle Q about XX = $(27)(1.5) = 40.5$ cm$^3$
From para. 4, $A\bar{y} = \Sigma ay = 135 + 40.5 = 175.5$ cm$^3$

Hence $\bar{y} = \dfrac{175.5}{A} = \dfrac{175.5}{45} = 3.9$ cm.

**Thus the centroid of the L-section lies at a point outside of the cross-section 3.1 cm from YY and 3.9 cm from XX.**

(When there are two parts forming a composite area, the centroid always lies on a straight line joining the centroids of the two separate parts—a fact which can be used as a check. In this problem, $C_P$, $C$ and $C_Q$ lie on the same straight line, as shown in *Fig 6*.)

*Problem 4* A metal gusset plate is as shown in *Fig 7*. Determine the position of its centroid.

The gusset place is divided into rectangle P, triangle Q and rectangle R, with their centroids $C_P$, $C_Q$ and $C_R$ respectively. When determining positions of centroids it is often more convenient to use a tabular approach as shown in *Table 1*.

TABLE 1

| Part | Area $a$ mm² | Distance of centroid from YY (i.e. (i.e. $x$ mm) | First moment of area about YY (i.e. $ax$ mm³) | Distance of centroid from XX (i.e. $y$ mm) | First moment of area about XX (i.e. $ay$ mm³) |
|---|---|---|---|---|---|
| P | 90 × 10 = 900 | 5 | (900)(5) = 4500 | $15 + \frac{1}{2}(90)$ = 60 | (900)(60) = 54 000 |
| Q | $\frac{1}{2}(72)(90)$ = 3240 | $10 + \frac{1}{3}(72)$ = 34 | (3240)(34) = 110 160 | $15 + \frac{1}{3}(90)$ = 45 | (3240)(45) = 145 800 |
| R | 82 × 15 = 1230 | 41 | (1230)(41) = 50 430 | 7.5 | (1230)(7.5) = 9225 |
| | $\Sigma a = A = 900 + 3240 + 1230 = 5370$ | | $\Sigma ax = 4500 + 110\ 160 + 50\ 430 = 165\ 090$ | | $\Sigma ay = 54\ 000 + 145\ 800 + 9225 = 209\ 025$ |

Let $\bar{x}$ and $\bar{y}$ be the distances of the centroid from YY and XX respectively.
From para. 4, $A\bar{x} = \Sigma ax$, from which
$$\bar{x} = \frac{\Sigma ax}{A} = \frac{165\ 090}{5370} = 30.7 \text{ mm}$$

$A\bar{y} = \Sigma ay$, from which
$$\bar{y} = \frac{\Sigma ay}{A} = \frac{209\ 025}{5370} = 38.9 \text{ mm}$$

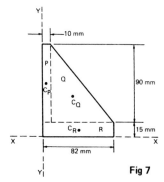

Fig 7

**Hence the centroid of the gusset plate lies at a point 30.7 mm from YY and 38.9 mm from XX**

71

*Problem 5* A rectangular template has dimensions of 30 cm by 20 cm. A 10 cm diameter hole is removed from the plate in the position shown in *Fig 8*. Determine the position of the centroid of the template.

The centroids of the rectangle and the circle to be removed are denoted by $C_R$ and $C_C$ respectively. In the *Table 2*, the area, and thus the first moment of area, of the circle is shown as negative since the circle is removed. Although the position of $\bar{x}$ can be seen by inspection, it is calculated in *Table 2* to illustrate the general method.

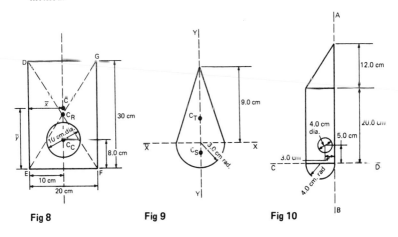

Fig 8      Fig 9      Fig 10

TABLE 2

| Part | Area $a$ cm² | Distance of centroid from DE (i.e. $x$ cm) | First moment of area about DE (i.e. $ax$ cm³) | Distance of centroid from EF (i.e. $y$ cm) | First moment of area about EF (i.e. $ay$ cm³) |
|---|---|---|---|---|---|
| Rectangle | 30 × 20 = 600 | 10 | (600)(10) = 6000 | 15 | (600)(15) = 9000 |
| Circle | $-\pi r^2 =$ $-\pi\left(\dfrac{10}{2}\right)^2$ $= -78.54$ | 10 | (−78.54)(10) = −785.4 | 8.0 | (−78.54)(8.0) = −628.32 |
| $\Sigma a = A = 600 - 78.54$ $= 521.46$ | | $\Sigma ax = 6000 - 785.4$ $= 5214.6$ | | $\Sigma ay = 9000 - 628.32$ $= 8371.68$ | |

If $\bar{x}$ and $\bar{y}$ are the distances of the centroid from DE and EF respectively then, from para 4,

$$A\bar{x} = \Sigma ax, \text{ from which } \bar{x} = \frac{\Sigma ax}{A} = \frac{5214.6}{521.46} = 10.0 \text{ cm}$$

and  $A\bar{y} = \Sigma ay$, from which $\bar{y} = \frac{\Sigma ay}{A} = \frac{8371.68}{521.46} = 16.1$ cm.

**Hence the centroid of the template lies at a point 10.0 cm from its left-hand edge and 16.1 cm from its bottom edge.**

(Note that $C_R$, $C_C$ and C lie on the same straight line.)

---

*Problem 6* Determine the position of the centroid of the area shown in *Fig 9*.

---

Let $C_T$ and $C_S$ be the centroids of the triangle and semicircle respectively. Since the area is symmetrical about its axis of symmetry YY, the centroid of the whole area will lie on YY. The horizontal axis XX has areas above and below it. In such cases, distances of centroids above XX are considered positive whilst distances of centroids below are considered negative, as shown in *Table 3*.

TABLE 3

| Part | Area $a$ cm$^2$ | Distance of centroid from XX (i.e. $y$ cm) | First moment of area about XX (i.e. $ay$ cm$^3$) |
|---|---|---|---|
| Triangle | $\frac{1}{2}(6.0)(9.0)$ $= 27.0$ | $\frac{1}{3}(9.0) = +3.0$ | $(27.0)(3.0)$ $= +81.0$ |
| Semicircle | $\frac{1}{2}\pi(3.0)^2$ $= 14.14$ | $\frac{-4(3.0)}{3\pi} = -1.27$ | $(14.14)(-1.27)$ $= -17.96$ |
| $\Sigma a = A = 41.14$ | | $\Sigma ay = +81.0 - 17.96 = +63.04$ | |

If $\bar{y}$ is the distance of the centroid from XX then,

from para. 4, $A\bar{y} = \Sigma ay$, from which $\bar{y} = \frac{\Sigma ay}{A} = \frac{+63.04}{41.14} = +1.53$ cm

**Hence the centroid of the area lies on the axis of symmetry YY, 1.53 cm above the axis XX.**

---

*Problem 7* Find the position of the centroid of the metal template shown in *Fig 10* about the right-hand edge and the diameter of the semicircle. The circular area is removed.

TABLE 4

| Part | Area $a$ cm³ | Distance of centroid from AB (i.e. $x$ cm) | First moment of area about AB (i.e. $ax$ cm³) | Distance of centroid from CD (i.e. $y$ cm) | First moment or area about CD (i.e. $ay$ cm³) |
|---|---|---|---|---|---|
| Triangle | $\frac{1}{2}$(8.0)(12.0) = 48.0 | $\frac{1}{3}$(8.0) = 2.667 | (48.0)(2.667) = 128.0 | $20.0 + \frac{1}{3}$(12.0) = 24.0 | (48.0)(24.0) = 1152.0 |
| Rectangle | (20.0)(8.0) = 160.0 | $\frac{1}{2}$(8.0) = 4.0 | (160.0)(4.0) = 640.0 | $\frac{1}{2}$(20.0) = 10.0 | (160.0)(10.0) = 1600.0 |
| Circle | $-\pi(2.0)^2$ = $-12.566$ (minus since circle is removed) | 3.0 | $(-12.566)(3.0)$ = $-37.70$ | 5.0 | $(-12.566)(5.0)$ = $-62.83$ |
| Semicircle | $\frac{1}{2}\pi(4.0)^2$ = 25.133 | 4.0 | (25.133)(4.0) = 100.5 | $\frac{-4(4.0)}{3\pi}$ = $-1.698$ (minus since below CD) | (25.133)($-1.698$) = $-42.68$ |
| $\Sigma a = A = 220.567$ | | $\Sigma ax = 830.8$ | | $\Sigma ay = 2646.49$ | |

If $\bar{x}$ and $\bar{y}$ are the distances of the centroid from AB and CD respectively then,

from para. 4, $A\bar{x} = \Sigma ax$, from which $\bar{x} = \dfrac{\Sigma ax}{A} = \dfrac{830.8}{220.567} = 3.77$ cm,

and $A\bar{y} = \Sigma ay$, from which $\bar{y} = \dfrac{\Sigma ay}{A} = \dfrac{2646.49}{220.567} = 12.0$ cm.

**Hence the centroid lies at a point 3.77 cm to the left of AB and 12.0 cm above CD.**

## C. FURTHER PROBLEMS ON CENTROIDS OF SIMPLE SHAPES

1   Determine the first moment of area of each of the shapes shown in *Fig 11* about axes XX and YY.

$$\begin{bmatrix} \text{(a) } 427.5 \text{ cm}^3, 247.5 \text{ cm}^3; \text{ (b) } 37\,700 \text{ mm}^3, 43\,980 \text{ mm}^3 \\ \text{(c) } 252 \text{ cm}^3, 252 \text{ cm}^3; \quad \text{(d) } 3808 \text{ mm}^3, 3927 \text{ mm}^3 \end{bmatrix}$$

2   Determine the distances from axes XX and YY of the centroid for the shapes shown in *Fig 12*.

$$\begin{bmatrix} \text{(a) } 16.3 \text{ cm below XX on YY;} \\ \text{(b) } 14.2 \text{ mm below XX and } 28.0 \text{ mm to the right of YY;} \\ \text{(c) } 6.66 \text{ cm above XX on YY.} \end{bmatrix}$$

**Fig 11**

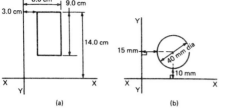
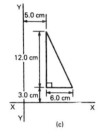
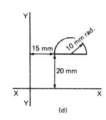
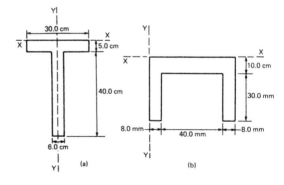
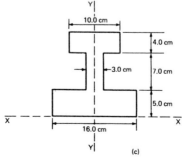

**Fig 12**

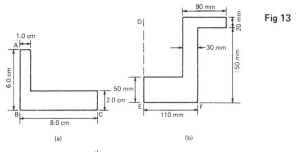

**Fig 13**

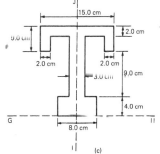

3. Find the positions of the centroids for the shapes shown in *Fig 13*.
   [(a) 2.30 cm to the right of AB, 1.60 cm above BC; (b) 78.9 mm to the right of DE, 70.4 mm above EF; (d) 10.1 cm above GH on IJ]

**Fig 14**

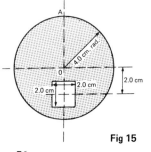

4. Determine the positions of the centroids for the templates shown in *Fig 14*. (In *Fig 14(a)*, the circular area is removed.)
   [(a) 56.0 mm from bottom edge, 44.3 mm from the left-hand edge; (b) 4.56 cm below the top edge, 3.56 cm from the right-hand edge.]

5. Determine the centroid of the shaded area in *Fig 15*.  [On OA, 0.173 cm from 0]

**Fig 15**

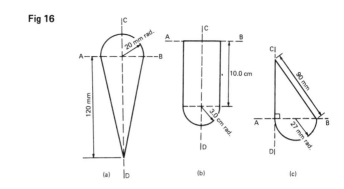

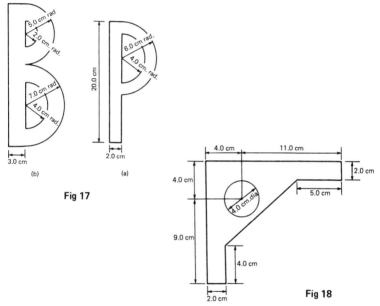

6. Find the positions of the centroids of the shapes shown in *Fig 16* about axes AB and CD.

$$\begin{bmatrix} \text{(a) 29.9 mm below AB on CD;} \\ \text{(b) 6.20 cm below AB on CD;} \\ \text{(c) 10.9 mm above AB, 21.3 mm to the right of CD.} \end{bmatrix}$$

7. Determine the positions of the centroids for the letters shown in *Fig 17*.

$$\begin{bmatrix} \text{(a) 11.8 cm from bottom, 2.86 cm from left-hand edge;} \\ \text{(b) 11.8 cm from bottom, 4.00 cm from left-hand edge;} \end{bmatrix}$$

8. In the template shown in *Fig 18*, the circular area is removed. Determine the position of the centroid of the template.

[3.95 cm from the top edge, 4.86 cm from the left-hand edge]

# 8 Further areas and volumes

## A. MAIN POINTS CONCERNED WITH AREAS AND VOLUMES

1. An **ellipse** is the name given to the regular oval shape PRQS shown in *Fig 1*. PQ is called the **major axis** and $a$ is the semi-major axis. RS is the **minor axis** and $b$ the semi-minor axis.
   **Area of ellipse PRQS** $= \pi ab$
   **Perimeter of ellipse PRQS** $= \pi(a+b) = \pi/2$ (sum of major and minor axes)
   (see *Problems 1 to 4*)

2. **Frustum and zone of a sphere**
   Volume of sphere $= 4/3\pi r^3$
   Surface area of sphere $= 4\pi r^2$
   A **frustum of a sphere** is the portion contained between two parallel planes. In *Fig 2*, PQRS is a frustum of the sphere. A **zone of a sphere** is the curved surface of a frustum. With reference to *Fig 2*:
   **Surface area of a zone of a sphere** $= 2\pi rh$
   **Volume of frustum of sphere** $= \dfrac{\pi h}{6}(h^2 + 3r_1^2 + 3r_2^2)$. (See *Problems 5 to 9*.)

3. **Prismoidal rule for finding volumes**
   The prismoidal rule applies to a solid of length $x$ divided by only three

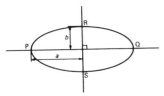

**Fig 1**

**Fig 2**

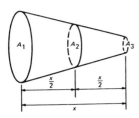

**Fig 3**

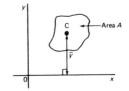

**Fig 4**

equidistant plane areas, $A_1$, $A_2$ and $A_3$ as shown in *Fig 3* and is merely an extension of Simpson's rule for volumes (see chapter 6, para. 2).

With reference to *Fig 3*: **Volume,** $V = \frac{x}{6}[A_1 + 4A_2 + A_3]$

The prismoidal rule gives precise values of volume for regular solids such as pyramids, cones, spheres and prismoids. (See *Problems 10 to 13*.)

4  A **theorem of Pappus** states:
'If a plane area is rotated about an axis in its own plane but not intersecting it, the volume of the solid formed is given by the product of the area and the distance moved by the centroid of the area.'
i.e., volume generated = area × distance moved through by the centroid.
In *Fig 4*, let C be the centroid of area $A$, and let $\overline{y}$ be the perpendicular distance of C from axis OX, then the distance moved by the centroid when area $A$ makes one revolution about OX is $2\pi\overline{y}$ (i.e. the circumference of a circle). Hence by Pappus' theorem:
Volume generated = area × $2\pi\overline{y}$, i.e. $V = 2\pi A\overline{y}$ cubic units.
(See *Problems 14 to 16*.)

## B. WORKED PROBLEMS ON AREAS AND VOLUMES

THE ELLIPSE

*Problem 1* The major axis of an ellipse is 12.0 cm and the minor axis is 7.0 cm. Determine the perimeter and the area of the ellipse.

From para. 1, the perimeter of an ellipse = $\pi(a+b)$, where $a$ and $b$ are the semi-major and semi-minor axes respectively.
Hence $a = 12.0/2 = 6.0$ cm and $b = 7.0/2 = 3.5$ cm.
Perimeter of ellipse = $\pi(a+b) = \pi(6.0+3.5) = $ **29.85 cm**
Area of ellipse = $\pi ab = \pi(6.0)(3.5) = $ **65.97 cm$^2$**

*Problem 2* Determine the total surface area and the volume of a regular elliptical cylinder, the diameters being 20.0 cm (major axis) and 15.0 cm (minor axis) respectively and the height 12.0 cm.

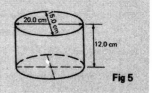

Fig 5

The cylinder is shown in *Fig 5*. Since the major axis is 20.0 cm the semi-major axis is $20.0/2 = 10.0$ cm, and since the minor axis is 15.0 cm the semi-minor axis is $15.0/2 = 7.5$ cm.

From para 1, area of elliptical end = $\pi ab = \pi(10.0)(7.5) = 75\pi$ cm² and the perimeter of elliptical end = $\pi(a+b) = \pi(10.0+7.5) = 17.5\pi$ cm.
Total surface area of cylinder = curved surface area + area of two ends
$$= (17.5\pi)(12.0) + 2(75\pi)$$
$$= 210\pi + 150\pi = 360\pi = 1131 \text{ cm}^2.$$
Volume of cylinder = area of base × perpendicular height
$$= (75\pi)(12.0) = 2827 \text{ cm}^3.$$

*Problem 3* A rectangular metal plate measures 16 cm by 10 cm. If the maximum possible sized ellipse is cut from the plate, determine the amount of metal wasted.

The plate is shown in *Fig 6*. Area of metal plate = 16 × 10 = 160 cm².
Area of ellipse, having semi-major axis, $a = 16/2 = 8$ cm and semi-minor axis, $b = 10/2 = 5$ cm, is given by:
Area = $\pi ab = \pi(8)(5) = 40\pi = 125.7$ cm².
Hence metal wasted = $160 - 125.7 = 34.3$ cm².

**Fig 6**

*Problem 4* An elliptical plot of land has an area of 150 m² and a circumference of 50 m. Determine the maximum length and maximum width of the plate.

Let the semi-major and semi-minor axes be $a$ and $b$ respectively (as in *Fig 1*).

Area of ellipse = $\pi ab = 150$     (1)
Circumference (or perimeter) of ellipse = $\pi(a+b) = 50$     (2)

From equation (2), $a+b = 50/\pi$, from which $a = (50/\pi) - b$
Substituting, $a = 50/\pi - b$ into equation (1) gives:
$\pi \left(\dfrac{50}{\pi} - b\right) b = 150$ i.e. $50b - \pi b^2 = 150$
from which, $\pi b^2 - 50b + 150 = 0$

Using the quadratic formula, $b = \dfrac{50 \pm \sqrt{[(-50)^2 - 4(\pi)(150)]}}{2\pi}$
$$= \dfrac{50 \pm 24.8}{2\pi} = 11.90 \text{ m or } 4.01 \text{ m}$$

Substituting in equation (1) gives: $\pi a(11.90) = 150$

i.e.      $a = \dfrac{150}{\pi(11.90)} = 4.01$ m    or    $\pi a(4.01) = 150$

i.e.      $a = \dfrac{150}{\pi(4.01)} = 11.91$ m.

Hence the semi-major and semi-minor axes are 11.91 m and 4.01 m. The maximum length of the plot = 2 × 11.91 = **23.82 m** and the maximum width of the plot = 2 × 4.01 = **8.02 m**

*Further problems on the ellipse may be found in section C, problems 1 to 6, page 87.*

## FRUSTUM AND ZONE OF A SPHERE

*Problem 5* Determine the volume of a frustum of a sphere of diameter 49.74 cm if the diameter of the ends of the frustum are 24.0 cm and 40.0 cm, and the height of the frustum is 7.00 cm.

From para. 2, Volume of frustum of a sphere $= \frac{\pi h}{6}(h^2 + 3r_1^2 + 3r_2^2)$,
where $h = 7.00$ cm, $r_1 = 24.0/2 = 12.0$ cm and $r_2 = 40.0/2 = 20.0$ cm.
Hence volume of frustum $= \frac{\pi(7.00)}{6}[(7.00)^2 + 3(12.0)^2 + 3(20.0)^2]$
$= \mathbf{6161 \ cm^3}$

*Problem 6* Determine for the frustum of *Problem 5* the curved surface area of the frustum.

The curved surface area of the frustum = surface area of zone = $2\pi rh$
(from para. 2), where $r$ = radius of sphere = $\frac{49.74}{2} = 24.87$ cm and $h = 7.00$ cm.
Hence, surface area of zone = $2\pi(24.87)(7.00) = \mathbf{1094 \ cm^2}$

*Problem 7* The diameters of the ends of the frustum of a sphere are 14.0 cm and 26.0 cm respectively and the thickness of the frustum is 5.0 cm. Determine, correct to 3 significant figures (a) the volume of the frustum of the sphere, (b) the radius of the sphere and (c) the area of the zone formed.

The frustum is shown shaded in the cross section of *Fig 7*.

Volume of frustum of sphere $= \frac{\pi h}{6}(h^2 + 3r_1^2 + 3r_2^2)$,
from para. 2, where $h = 5.0$, $r_1 = \frac{14.0}{2} = 7.0$ cm
and $r_2 = \frac{26.0}{2} = 13.0$ cm

Hence volume of frustum of sphere
$= \frac{\pi(5.0)}{6}[(5.0)^2 + 3(7.0)^2 + 3(13.0)^2]$
$= \frac{\pi(5.0)}{6}[25.0 + 147.0 + 507.0]$
$= \mathbf{1780 \ cm^3}$, correct to 3 significant figures

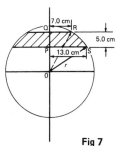

**Fig 7**

(b) The radius, $r$, of the sphere may be calculated using *Fig 7*.
Using Pythagoras' theorem $OS^2 = PS^2 + OP^2$, i.e. $r^2 = (13.0)^2 + OP^2$     (1)
Also      $OR^2 = QR^2 + OQ^2$, i.e. $r^2 = (7.0)^2 + OQ^2$
However $OQ = QP + OP = 5.0 + OP$
Therefore $r^2 = (7.0)^2 + (5.0 + OP)^2$     (2)
Equating equations (1) and (2) gives:   $(13.0)^2 + OP^2 = (7.0)^2 + (5.0 + OP)^2$
$$169.0 + OP^2 = 49.0 + 25.0 + 10.0(OP) + OP^2$$
$$169.0 = 74.0 + 10.0(OP)$$

Hence      $OP = \dfrac{16.90 - 74.0}{10.0} = 9.50$ cm

Substituting $OP = 9.50$ cm into equation (1) gives $r^2 = (13.0)^2 + (9.50)^2$
from which $r = \sqrt{[(13.0)^2 + (9.50)^2]}$
i.e. **radius of sphere, $r = 16.1$ cm**.

(c) Area of zone of sphere $= 2\pi r h = 2\pi(16.1)(5.0)$
$$= 506 \text{ cm}^2, \text{ correct to 3 significant figures.}$$

---

*Problem 9* A frustum of a sphere of diameter 12.0 cm is formed by two parallel planes, one through the diameter and the other distance $h$ from the diameter. The curved surface area of the frustum is required to be 1/4 of the total surface area of the sphere. Determine (a) the volume and surface area of the sphere, (b) the thickness $h$ of the frustum, (c) the volume of the frustum and (d) the volume of the frustum expressed as a percentage of the sphere.

(a) Volume of sphere $V = \dfrac{4}{3}\pi r^3 = \dfrac{4}{3}\pi(12.0/2)^3 = \mathbf{904.8 \text{ cm}^3}$

Surface area of sphere $= 4\pi r^2 = 4\pi(12.0/2)^2 = \mathbf{452.4 \text{ cm}^2}$

(b) Curved surface area of frustum $= \dfrac{1}{4} \times$ surface area of sphere
$$= \dfrac{1}{4} \times 452.4 = 113.1 \text{ cm}^2$$
From para. 2, $113.1 = 2\pi r h = 2\pi(12.0/2)h$

Hence thickness of frustum, $h = \dfrac{113.1}{2\pi(6.0)} = \mathbf{3.0 \text{ cm}}$

(c) Volume of frustum, $V = \dfrac{\pi h}{6}(h^2 + 3r_1^2 + 3r_2^2)$,

where $h = 3.0$ cm, $r_2 = 6.0$ cm and
$r_1 = \sqrt{(OQ^2 - OP^2)}$, from *Fig 8*, i.e.
$r_1 = \sqrt{[(6.0)^2 - (3.0)^2]} = 5.196$ cm.
Hence volume of frustum
$$= \dfrac{\pi(3.0)}{6}[(3.0)^2 + 3(5.196)^2 + 3(6.0)^2]$$
$$= \dfrac{\pi}{2}[9.0 + 81 + 108.0] = \mathbf{311.0 \text{ cm}^3}$$

(d) $\dfrac{\text{Volume of frustum}}{\text{Volume of sphere}} = \dfrac{311.0}{904.8} \times 100\%$
$$= \mathbf{34.37\%}.$$

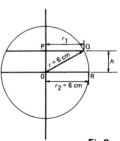

Fig 8

*Problem 9* A spherical storage tank is filled with liquid to a depth of 20 cm. If the internal diameter of the vessel is 30 cm, determine the number of litres of liquid in the container (1 litre = 1000 cm$^3$).

The liquid is represented by the shaded area in the section shown in *Fig 9*. The volume of liquid comprises a hemisphere and a frustum of thickness 5 cm.

Hence volume of liquid = $\frac{2}{3}\pi r^3 + \frac{\pi h}{6}[(h^2 + 3(r_1)^2 + 3(r_2)^2]$

where $r_2 = 30/2 = 15$ cm and $r_1 = \sqrt{[(15)^2 - (5)^2]} = 14.14$ cm.

Volume of liquid = $\frac{2}{3}\pi(15)^3 + \frac{\pi(5)}{6}[5^2 + 3(14.14)^2 + 3(15)^2]$
= 7069 + 3403
= 10 470 cm$^3$.

Since 1 litre = 1000 cm$^3$, the number of litres of liquid = $\frac{10\,470}{1000}$ = **10.47 litres**

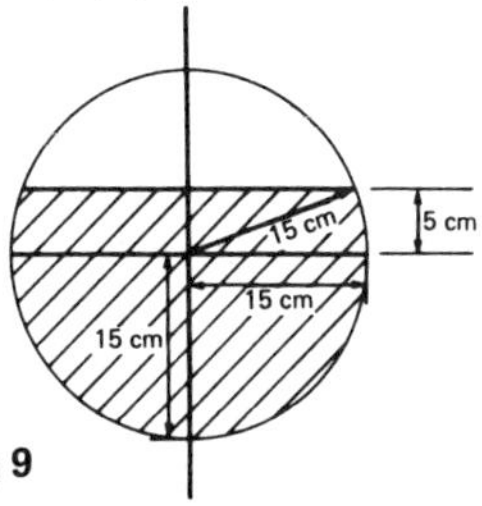

**Fig 9**

*Further problems on frustums and zones of spheres may be found in section C, Problems 7 to 11, page 88.*

PRISMOIDAL RULE

*Problem 10* A container is in the shape of a frustum of a cone. Its diameter at the bottom is 18 cm and at the top 30 cm. If the depth is 24 cm determine the capacity of the container, correct to the nearest litre by the Prismoidal rule (1 litre = 1000 cm$^3$).

The container is shown in *Fig 10*. At the midpoint, i.e. at a distance of 12 cm from one end the radius $r_2$ is $(9+15)/2 = 12$ cm, since the sloping sides change uniformly.
Volume of container by the prismoidal rule
= $(x/6)[A_1 + 4A_2 + A_3]$, from para. 3, where
$x = 24$ cm, $A_1 = \pi(15)^2$ cm$^2$, $A_2 = \pi(12)^2$ cm$^2$
and $A_3 = \pi(9)^2$ cm$^2$.

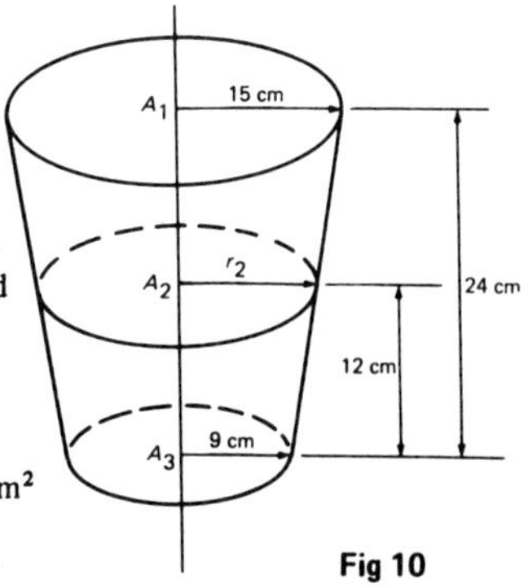

**Fig 10**

Hence volume of container = $\frac{24}{6} [\pi(15)^2 + 4\pi(12)^2 + \pi(9)^2]$

$= 4[706.86 + 1809.56 + 254.47]$

$= 11\,080$ cm$^3 = \frac{11\,080}{1000}$ litres

$= \mathbf{11}$ **litres, correct to the nearest litre.**

[*Check*: Volume of frustum of cone $= \frac{1}{3}\pi h[R^2 + Rr + r^2]$ from chapter 5

$= \frac{1}{3}\pi(24)[(15)^2 + (15)(9) + (9)^2]$

$= 11\,080$ cm$^3$ as shown above]

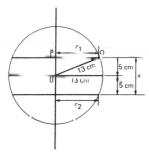

**Fig 11**

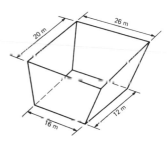

**Fig 12**

*Problem 11* A frustum of a sphere of radius 13 cm is formed by two parallel planes on opposite sides of the centre, each at distances of 5 cm from the centre. Determine the volume of the frustum (a) by using the prismoidal rule, and (b) by using the formula for the volume of a frustum of a sphere.

The frustum of the sphere is shown by the section in *Fig 11*.
Radius $r_1 = r_2 = PQ = \sqrt{(13^2 - 5^2)} = 12$ cm, by Pythagoras' theorem.

(a) Using the prismoidal rule, volume of frustum, $V = \frac{x}{6}[A_1 + 4A_2 + A_3]$

$= \frac{10}{6}[\pi(12)^2 + 4\pi(13)^2 + \pi(12)^2]$

$= \frac{10\pi}{6}[144 + 676 + 144]$

$= \mathbf{5047}$ **cm**$^3$

(b) Using the formula for the volume of a frustum of a sphere (from para. 2),

Volume $V = \frac{\pi h}{6}(h^2 + 3r_1^2 + 3r_2^2) = \frac{\pi(10)}{6}[10^2 + 3(12)^2 + 3(12)^2]$

$= \frac{10\pi}{6}(100 + 432 + 432) = \mathbf{5047}$ **cm**$^3$

*Problem 12* A hole is to be excavated in the form of a prismoid. The bottom is to be a rectangle 16 m long by 12 m wide; the top is also a rectangle, 26 m long by 20 m wide. Find the volume of earth to be removed, correct to 3 significant figures, if the depth of the hole is 6.0 m.

The hole is shown in *Fig 12*. Let $A_1$ represent the area of the top of the hole, i.e. $A_1 = 20 \times 26 = 520$ m$^2$.
Let $A_3$ represent the area of the bottom of the hole, i.e. $A_3 = 16 \times 12 = 192$ m$^2$
Let $A_2$ represent the rectangular area through the middle of the hole parallel to areas $A_1$ and $A_3$. The length of this rectangle is $(26+16)/2 = 21$ m and the width is $(20+12)/2 = 16$ m, assuming the sloping edges are uniform.
Thus area $A_2 = 21 \times 16 = 336$ m$^2$.

Using the prismoidal rule, volume of hole $= \frac{x}{6}[A_1 + 4A_2 + A_3]$

$$= \frac{6}{6}[520 + 4(336) + 192]$$

$$= 2056 \text{ m}^3 = \mathbf{2060 \text{ m}^3}, \text{ correct to 3 significant figures.}$$

*Problem 13* The roof of a building is in the form of a frustum of a pyramid with a square base of side 5.0 m. The flat top is a square of side 1.0 m and all the sloping sides are pitched at the same angle. The vertical height of the flat top above the level of the eaves is 4.0m. Calculate, using the prismoidal rule, the volume enclosed by the roof.

Let area of top of frustum be $A_1 = (1.0)^2 = 1.0$ m$^2$
Let area of bottom of frustum be $A_3 = (5.0)^2 = 25.0$ m$^2$
Let area of section through the middle of the frustum parallel to $A_1$ and $A_3$ be $A_2$. The length of the side of the square forming $A_2$ is the average of the sides forming $A_1$ and $A_3$, i.e. $(1.0+5.0)/2 = 3.0$ m. Hence $A_2 = (3.0)^2 = 9.0$ m$^2$.

Using the prismoidal rule, volume of frustum $= \frac{x}{6}[A_1 + 4A_2 + A_3]$

$$= \frac{4.0}{6}[1.0 + 4(9.0) + 25.0]$$

**Hence volume enclosed by roof = 41.3 m$^3$**

*Further problems on the prismoidal rule may be found in section C, Problems 12 to 15, page 88.*

### THEOREM OF PAPPUS

*Problem 14* Use the theorem of Pappus to calculate the position of the centroid of a semicircle of radius $r$.

If the semicircular area shown in *Fig 13* is rotated about axis XX, then the volume of the solid generated is that of a sphere of volume $4/3\pi r^3$. The area of the semicircle is $\pi r^2/2$. Centroid C is shown in *Fig 13* on the axis of symmetry OY. By the theorem of Pappus, volume generated = area × distance moved through by centroid

i.e. $\quad \dfrac{4}{3}\pi r^3 = \left(\dfrac{\pi r^2}{2}\right)(2\pi \bar{y})$

Hence $\quad \bar{y} = \dfrac{\frac{4}{3}\pi r^3}{\left(\dfrac{\pi r^2}{2}\right)(2\pi)} = \dfrac{4r}{3\pi}$ units

i.e. the centroid of a semicircle lies on the axis of symmetry at a distance of $4r/3\pi$ from the diameter (see chapter 7).

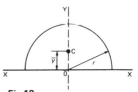

**Fig 13**

**Problem 15** Determine the volume of an anchor ring formed by rotating a circle of radius 6.0 cm about an axis at a distance of 25 cm from its centre. Give answer in cubic metres, correct to 3 significant figures.

The 6.0 cm radius circle is rotated about axis XX as shown by the side elevation in *Fig 14*.
Area of circle = $\pi(6.0)^2 = 36.0\pi$ cm$^2$.
Distance moved by centroid C in one complete revolution = $2\pi(25) = 50\pi$ cm (i.e. circumference of a circle of radius 25 cm).
By the theorem of Pappus, volume generated
= area × distance moved by centroid, i.e., volume of anchor ring = $(36.0\pi)(50\pi) = 17\,770$ cm$^3$

$= \dfrac{17\,770}{10^6}$ m$^3$ = 0.0178 m$^3$

correct to 3 significant figures.

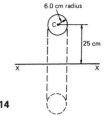

**Fig 14**

**Problem 16** A metal disc has a radius of 5.0 cm and is of thickness 2.0 cm. A semicircular groove of diameter 2.0 cm is machined centrally around the rim to form a pulley. Determine, using Pappus' theorem, the volume of metal removed and the volume of the pulley.

A side view of the rim of the disc is shown in *Fig 15*. When area PQRS is rotated about axis XX the volume generated is that of the pulley. The centroid of the semicircular area removed is at a distance $4r/3\pi$ from its diameter (see *Problem 14*), i.e $4(1.0)/3\pi$, i.e. 0.424 cm from PQ. Thus the distance of the centroid from XX is 5.0−0.424, i.e. 4.576 cm. The distance moved through in one revolution by the centroid is $2\pi(4.576)$ cm.

Area of semicircle = $\frac{\pi r^2}{2} = \frac{\pi(1.0)^2}{2} = \frac{\pi}{2}$ cm$^2$.

By the theorem of Pappus, volume generated = area × distance moved by centroid

$$= \left(\frac{\pi}{2}\right)(2\pi 4.576)$$

i.e. **volume of metal removed = 45.16 cm$^3$**
**Volume of pulley** = volume of cylindrical disc
  −volume of metal removed
 = $\pi(5.0)^2(2.0) - 45.16$
 = **111.9 cm$^3$**

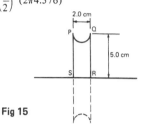

**Fig 15**

*Further problems on the theorem of Pappus may be found in the following section C, Problems 16 to 21, page 88.*

---

## C. FURTHER PROBLEMS ON AREAS AND VOLUMES

THE ELLIPSE

1  The major axis of an ellipse is 200 mm and the minor axis is 125 mm. Determine the perimeter and the area of the ellipse.

[510.5 mm; 19 630 mm$^2$]

2  A regular elliptical closed cylinder has diameters of 12.0 cm (major axis) and 8.0 cm (minor axis) and the perpendicular height is 15.0 cm. Determine the total surface area and the volume of the cylinder.

[622 cm$^2$; 1131 cm$^3$]

3  An elliptical fish pond has an area of 120 m$^2$. If its greatest length is 25.0 m, find the perimeter of the pond.

[48.87 m]

4  Determine the cost, correct to the nearest pound, of enclosing an elliptical plot of land, having major and minor diameter lengths of 80.2 m and 30.0 m, with fencing costing £8 per metre length.

[£1385]

5  An ellipse has an area of 125 cm$^2$ and a perimeter of 64 cm. Determine the maximum length and maximum breadth of the ellipse.

[36.36 cm; 4.38 cm]

6  A greyhound track is in the form of an ellipse, the axes being 160 m and 100 m respectively for the inner boundary and 175 m and 115 m for the outer boundary. Calculate the area of the track.

[3240 m$^2$]

FRUSTUM AND ZONE OF A SPHERE

7 Determine the volume and surface area of a frustum of a sphere of diameter 47.85 cm if the radii of the ends of the frustum are 14.0 cm and 22.0 cm and the height of the frustum is 10.0 cm.

[11 210 cm$^3$ ; 1503 cm$^2$]

8 Determine the volume (in cm$^3$) and the surface area (in cm$^2$) of a frustum of a sphere if the diameter of the ends are 80.0 mm and 120.0 mm and the thickness is 30.0 mm.

[259.2 cm$^3$ ; 118.3 cm$^2$]

9 A sphere has a radius of 6.50 cm. Determine its volume and surface area. A frustum of the sphere is formed by two parallel planes, one through the diameter and the other at a distance $h$ from the diameter. If the curved surface area of the frustum is to be 1/5 of the surface area of the sphere, find the height $h$ and the volume of the frustum.

[1150 cm$^3$ ; 531 cm$^2$ ; 2.60 cm; 326.7 cm$^3$]

10 A sphere has a diameter of 32.0 mm. Calculate the volume (in cm$^3$) of the frustum of the sphere contained between two parallel planes distances 12.0 mm and 10.0 mm from the centre and on opposite sides of it.

[14.84 cm$^3$]

11 A spherical storage tank is filled with liquid to a depth of 30.0 cm. If the inner diameter of the vessel is 45.0 cm determine the number of litres of liquid in the container (1 litre = 1000 cm$^3$).

[35.34 l]

PRISMOIDAL RULE

12 Use the prismoidal rule to find the volume of a frustum of a sphere contained between two parallel planes on opposite sides of the centre each of radius 7.0 cm and each 4.0 cm from the centre.

[1500 cm$^3$]

13 Determine the volume of a cone of perpendicular height 16.0 cm and base diameter 10.0 cm by using the prismoidal rule.

[418.9 cm$^3$]

14 A bucket is in the form of a frustum of a cone. The diameter of the base is 28.0 cm and the diameter of the top is 42.0 cm. If the depth is 32.0 cm, determine the capacity of the bucket (in litres) using the prismoidal rule (1 litre = 1000 cm$^3$).

[31.20 litre]

15 Determine the capacity of a water reservoir, in litres, the top being a 30.0 m by 12.0 m rectangle, the bottom being a 20.0 m by 8.0 m rectangle and the depth being 5.0 m. (1 litre = 1000 cm$^3$).

[1.267 × 10$^6$ litre]

THEOREM OF PAPPUS

16 A right angled isosceles triangle having a hypotenuse of 8 cm is revolved one revolution about one of its equal sides as axis. Determine the volume of the solid generated using Pappus' theorem.

[189.6 cm$^3$]

17  A rectangle measuring 10.0 cm by 6.0 cm rotates one revolution about one of its longest sides as axis. Determine the volume of the resulting cylinder by using the theorem of Pappus.

[1131 cm$^2$]

18  Determine the volume of a metal ring formed by rotating a circle of radius 9.0 cm about an axis at a perpendicular distance of 20.0 cm from its centre.

[31 980 cm$^3$]

19  A steel ring is formed by rotating an ellipse having major and minor axes of 10.0 cm and 7.0 cm about an axis at a perpendicular distance of 30.0 cm from its centre. Determine the volume of the ring using the theorem of Pappus.

[10 360 cm$^3$]

20  A metal disc has a radius of 7.0 cm and is of thickness 2.5 cm. A semi-circular groove of diameter 2.0 cm is machined centrally around the rim to form a pulley. Determine the volume of metal removed using Pappus' theorem and express this as a percentage of the original volume of the disc.

[64.90 cm$^3$; 16.86%]

21  One edge of an equilateral triangle of side 10.0 cm is parallel to an axis XX lying at a distance of 12.0 cm from the edge and the apex is on the opposite side to the axis. If the triangle is rotated 360° about axis XX, determine the volume of the solid generated.

[4050 cm$^3$]

# 9 Presentation of grouped data

## A. MAIN POINTS CONCERNED WITH PRESENTATION OF GROUPED DATA

1 Data is obtained largely by two methods:
   (a) by counting, for example, the number of stamps sold by a post office in equal periods of time, and
   (b) by measurement, for example, the heights of a group of people.
2 When data is obtained by counting, only whole numbers are possible and the data is called **discrete**. Measured data can have any value within certain limits and is called **continuous** (see *Problem 1*).
3 A **set** is a group of data and an individual value within the set is called a **member** of the set. Thus, if the masses of five people are measured, correct to the nearest 0.1 kilogram and are found to be 53.1 kg, 59.4 kg, 62.1 kg, 77.8 kg and 64.4 kg, then the set of masses in kilograms for the five people is

{ 53.1, 59.4, 62.1, 77.8, 64.4 }

and one of the members of the set is 59.4.
4 The number of times that the same value of a member occurs in a set is called the **frequency** of that member. Thus in the set 2, 3, 4, 5, 4, 2, 4, 7, 9, member 4 has a frequency of three, member 2 has a frequency of two and the other members have a frequency of one.
5 When the number of members in a set is small, say ten or less, the data can be represented diagrammatically without further analysis, by means of pictograms, bar charts, percentage components bar charts of pie diagrams. (See *Mathematics I Checkbook*). For sets having more than ten members, those members having similar values are grouped together in **classes** to form a **frequency distribution**.

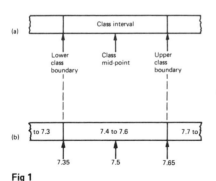

**Fig 1**

A frequency distribution is a table showing classes and their corresponding frequencies, (see Problems 2 and 4). The new set of data obtained by forming a frequency distribution is called **grouped data**.

6 The terms used in connection with grouped data are shown in *Fig 1(a)*. The size or range of a class is given by the **upper class boundary value** minus the **lower class boundary value**,

and in *Fig 1* is 7.65−7.35, i.e., 0.3. The **class interval** for the class shown in *Fig 1* is 7.4 to 7.6 and the class mid-point value is given by

$$\frac{\text{upper class boundary value} - \text{lower class boundary value}}{2},$$

and in *Fig 1* is $\frac{7.65-7.35}{2}$, i.e., 7.5.

7   One of the principal ways of presenting grouped data diagrammatically is by using a **histogram**, in which the **areas** of vertical, adjacent rectangles are made proportional to frequencies of the classes (see *Problem 3*). When class intervals are equal, the heights of the rectangles of a histogram are equal to the frequencies of the classes. For histograms having unequal class intervals, the area must be proportional to the frequency. Hence, if the class interval of class A is twice the class interval of class B, then for equal frequencies, the height of the rectangle representing A is half that of B (see *Problem 5*).

8   A **cumulative frequency distribution** is a table showing the cumulative frequency for each value of upper class boundary. The cumulative frequency for a particular value of upper class boundary is obtained by adding the frequency of the class to the sum of the previous frequencies. A cumulative frequency distribution is formed in *Problem 6*.

9   The curve obtained by joining the co-ordinates of cumulative frequency, (vertically), against upper class boundary (horizontally) is called an **ogive** or a **cumulative frequency distribution curve** (see *Problem 7*).

## B. WORKED PROBLEMS ON THE PRESENTATION OF GROUPED DATA

*Problem 1* Data is obtained on the topics given below. State whether it is discrete or continuous data.
(a) The number of days on which rain falls in a month for each month of the year.
(b) The mileage travelled by each of a number of salesmen.
(c) The time that each of a batch of similar batteries lasts.
(d) The amount of money spent by each of several families on food.

(a) The number of days on which rain falls in a given month must be an integer value and is obtained by **counting** the number of days. Hence, this data is **discrete**.
(b) A salesman can travel any number of miles (and parts of a mile) between certain limits and this data is **measured**. Hence the data is **continuous**.
(c) The time that a battery lasts is **measured** and can have any value between certain limits. Hence this data is **continuous**.
(d) The amount of money spent on food can only be expressed correct to the nearest half pence, the amount being **counted**. Hence, this data is **discrete**.

*Problem 2* The data given below refers to the gain of each of a batch of 40 transistors, expressed correct to the nearest whole number. Form a frequency distribution for this data having seven classes.

| | | | | | | | |
|---|---|---|---|---|---|---|---|
| 81 | 83 | 87 | 74 | 76 | 89 | 82 | 84 |
| 86 | 76 | 77 | 71 | 86 | 85 | 87 | 88 |
| 84 | 81 | 80 | 81 | 73 | 89 | 82 | 79 |
| 81 | 79 | 78 | 80 | 85 | 77 | 84 | 78 |
| 83 | 79 | 80 | 83 | 82 | 79 | 80 | 77 |

The **range** of the data is the value obtained by taking the value of the largest member from that of the smallest member. Inspection of the set of data shows that, range = 89 − 71 = 18. The size of each class is given approximately by range divided by the number of classes. Since 7 classes are required, the size of each class is 18/7, that is, approximately 3. To achieve seven equal classes spanning a range of values from 71 to 89, the class intervals are selected as: 70–72, 73–75, and so on.

To assist with accurately determining the number in each class, a **tally diagram** is produced, as shown in *Table 1(a)*. This is obtained by listing the classes in the

TABLE 1(a)

| Class | Tally |
|---|---|
| 70–72 | 1 |
| 73–75 | 11 |
| 76–78 | 1111 11 |
| 79–81 | 1111 1111 11 |
| 82–84 | 1111 1111 |
| 85–87 | 1111 1 |
| 88–90 | 111 |

TABLE 1(b)

| Class | Class mid-point | Frequency |
|---|---|---|
| 70–72 | 71 | 1 |
| 73–75 | 74 | 2 |
| 76–78 | 77 | 7 |
| 79–81 | 80 | 12 |
| 82–84 | 83 | 9 |
| 85–87 | 86 | 6 |
| 88–90 | 89 | 3 |

left-hand column, and then inspecting each of the 40 members of the set in turn and allocating them to the appropriate classes by putting "1's" in the appropriate rows. Every fifth '1' allocated to a particular row is shown as an oblique line crossing the four previous '1's, to help with final counting.

A frequency distribution for the data is shown in *Table 1(b)* and lists classes and their corresponding frequencies, obtained from the tally diagram. (Class mid-point values are also shown in the table, since they are used for constructing the histogram for this data (see *Problem 3*)).

*Problem 3* Construct a histogram for the data given in *Table 1*.

The histogram is shown in *Fig 2*. The width of the rectangles correspond to the upper class boundary values minus the lower class boundary values and the heights of the rectangles correspond to the class frequencies. The easiest way to draw a histogram is to mark the class mid-point values on the horizontal scale and draw the rectangles symmetrically about the appropriate class mid-point values and touching one another.

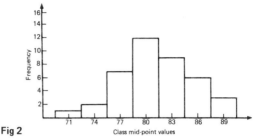

**Fig 2**

*Problem 4* The amount of money earned weekly by 40 people working in a factory, correct to the nearest £10, is shown below. Form a frequency distribution having 6 classes for this data.

| 80 | 90 | 70 | 110 | 90 | 160 | 110 | 80 |
| 140 | 30 | 90 | 50 | 100 | 110 | 60 | 100 |
| 80 | 90 | 110 | 80 | 100 | 90 | 120 | 70 |
| 130 | 170 | 80 | 120 | 100 | 110 | 40 | 110 |
| 50 | 100 | 110 | 90 | 100 | 70 | 110 | 80 |

Inspection of the set given shows that the majority of the members of the set lie between £80 and £110 and that there are a much smaller number of extreme values ranging from £30 to £180. If equal class intervals are selected, the frequency distribution obtained does not give as much information as one with unequal class intervals. Since the majority of members are between £80 and £100, the class intervals in this range are selected to be smaller than those outside of this range. There is no unique solution and one possible solution is shown in *Table 2*.

TABLE 2

| Class | Frequency |
|---|---|
| 20–40 | 2 |
| 50–70 | 6 |
| 80–90 | 12 |
| 100–110 | 14 |
| 120–140 | 4 |
| 150–170 | 2 |

*Problem 5* Draw a histogram for the data given in *Table 2*.

When dealing with unequal class intervals, the histogram must be drawn so that the **areas** (and not the heights of the rectangles), are proportional to the frequencies of the classes. The data given is shown in columns 1 and 2 of *Table 3*. Columns 3 and 4 give the upper and lower class boundaries respectively. In column 5,

TABLE 3

| 1<br>Class | 2<br>Frequency | 3<br>Upper class boundary | 4<br>Lower class boundary | 5<br>Class range | 6<br>Height of rectangle |
|---|---|---|---|---|---|
| 20–40 | 2 | 45 | 15 | 30 | $\frac{2}{30} = \frac{1}{15}$ |
| 50–70 | 6 | 75 | 45 | 30 | $\frac{6}{30} = \frac{3}{15}$ |
| 80–90 | 12 | 95 | 75 | 20 | $\frac{12}{20} = \frac{9}{15}$ |
| 100–110 | 14 | 115 | 95 | 20 | $\frac{14}{20} = \frac{10½}{15}$ |
| 120–140 | 4 | 145 | 115 | 30 | $\frac{4}{30} = \frac{2}{15}$ |
| 150–170 | 2 | 175 | 145 | 30 | $\frac{2}{30} = \frac{1}{15}$ |

the class ranges (i.e. upper class boundary minus lower class boundary values), are listed. The heights of the rectangles are proportional to the ratio (frequency/class range), as shown in column 6. The histogram is shown in *Fig 3*.

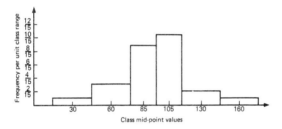

**Fig 3**

*Problem 6* The frequency distribution of the diameters of 60 bolts measured in millimetres is as follows:

3.93–3.94   1,   3.95–3.96   5,   3.97–3.98   9,
3.99–4.00   17,   4.01–4.02   15,   4.03–4.04   7,
4.05–4.06   4,   4.07–4.08   2.

Form a cumulative frequency distribution for this data.

A cumulative frequency distribution is a table listing values of upper class boundaries and cumulative frequencies of less than these upper class boundary values. Such a cumulative frequency distribution for the data given is shown in *Table 4*. Columns 1 and 2 show the classes and their frequencies. Column 3 lists the upper class boundary values. In column 4, the sum of the frequencies less than the upper class boundary values given in column 3 are listed. Thus, for the 3.97–3.98 class shown in row 3, the cumulative frequency is the sum of all frequencies less than 3.985, i.e., 1+5+9 = 15, and so on.

*Problem 7* Draw an ogive (cumulative frequency distribution curve), for the data given in *Table 4*.

### TABLE 4

| 1<br>Class | 2<br>Frequency | 3<br>Upper<br>class boundary | 4<br>Cumulative<br>frequency |
|---|---|---|---|
| 3.93–3.94 | 1 | less than 3.945 | 1 |
| 3.95–3.96 | 5 | less than 3.965 | 6 |
| 3.97–3.98 | 9 | less than 3.985 | 15 |
| 3.99–4.00 | 17 | less than 4.005 | 32 |
| 4.01–4.02 | 15 | less than 4.025 | 47 |
| 4.03–4.04 | 7 | less than 4.045 | 54 |
| 4.05–4.06 | 4 | less than 4.065 | 58 |
| 4.07–4.08 | 2 | less than 4.085 | 60 |

The ogive for the cumulative frequency distribution given in *Table 4* is shown in *Fig 4*. The co-ordinates corresponding to each upper class boundary/cumulative frequency value are plotted, and the co-ordinates are joined by straight lines. (*Note*: it is not the best curve drawn through the co-ordinates as is usual practice when drawing graphs of experimental results.) The ogive is 'anchored' at its start by assuming a zero cumulative frequency for a class one less than the first one given in *Table 4*, i.e., (3.925, 0).

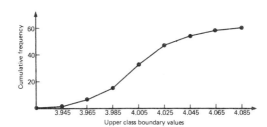

**Fig 4**

## C. FURTHER PROBLEMS ON THE PRESENTATION OF GROUPED DATA

In *Problems 1 and 2*, state whether data relating to the topics given is discrete or continuous.
1. (a) The amount of petrol produced daily, for each of 31 days, by a refinery.
   (b) The amount of coal produced daily by each of 15 miners.

    (c) The number of bottles of milk delivered daily by each of 20 milkmen.

    (d) The size of 10 samples of rivets produced by a machine.

                    [(a) Continuous; (b) continuous; (c) discrete; (d) continuous]

2  (a) The number of people visiting an exhibition on each of 5 days.

    (b) The time taken by each of 12 athletes to run 100 metres.

    (c) The value of stamps sold in a day by each of 20 post offices.

    (d) The number of defective items produced in each of 10 one-hour periods by a machine.

                    [(a) Discrete; (b) continuous; (c) discrete; (d) discrete]

3  The number of cars passing across a particular road junction is counted daily from Monday to Friday for a ten week period and the results are shown below. Form a frequency distribution of about 7 classes for this data.

| 86 | 87 | 78 | 78 | 82 | 85 | 88 | 71 | 74 | 80 |
|---|---|---|---|---|---|---|---|---|---|
| 77 | 73 | 84 | 79 | 77 | 74 | 81 | 91 | 85 | 82 |
| 79 | 86 | 72 | 79 | 74 | 76 | 80 | 85 | 83 | 84 |
| 85 | 88 | 81 | 77 | 90 | 84 | 81 | 84 | 85 | 83 |
| 83 | 82 | 81 | 81 | 78 | 80 | 82 | 75 | 87 | 80 |

[There is no unique solution, but one solution is: 71–73 3; 74–76 5; 77–79 9; 80–82 14; 83–85 11; 86–88 6; 89–91 2.]

4  The mass in kilograms, correct to the nearest one-tenth of a kilogram, of 60 bars of metal are as shown. Form a frequency distribution of about 8 classes for this data.

| 39.8 | 40.1 | 40.3 | 40.0 | 40.6 | 39.7 | 40.0 | 40.4 | 39.6 | 39.3 |
|---|---|---|---|---|---|---|---|---|---|
| 39.6 | 40.7 | 40.2 | 39.9 | 40.3 | 40.2 | 40.4 | 39.9 | 39.8 | 40.0 |
| 40.2 | 40.1 | 40.3 | 39.7 | 39.9 | 40.5 | 39.9 | 40.5 | 40.0 | 39.9 |
| 40.1 | 40.8 | 40.0 | 40.0 | 40.1 | 40.2 | 40.1 | 40.0 | 40.2 | 39.9 |
| 39.7 | 39.8 | 40.4 | 39.7 | 39.9 | 39.5 | 40.1 | 40.1 | 39.9 | 40.2 |
| 39.5 | 40.6 | 40.0 | 40.1 | 39.8 | 39.7 | 39.5 | 40.2 | 39.9 | 40.3 |

[There is no unique solution, but one solution is: 39.3–39.4 1; 39.5–39.6 5; 39.7–39.8 9; 39.9–40.0 17; 40.1–40.2 15; 40.3–40.4 7; 40.5–40.6 4; 40.7–40.8 2.]

5  Draw a histogram for the frequency distribution given in the solution of *Problem 3*.

[Rectangles, touching one another, having mid-points of 72, 75, 78, 81, ...... and heights proportional to 3, 5, 9, 14, ......]

6  Draw a histogram for the frequency distribution given in the solution of *Problem 4*.

[Rectangles, touching one another, having mid-points of 39.35, 39.55, 39.75, 39.95, ...... and heights of 1, 5, 9, 17, ......]

7  The force in kilonewtons, expressed correct to the nearest kilonewton, required to cause each of 40 blocks of material to fail are as shown. Produce a frequency distribution having about 6 classes for this data. (The frequency distribution should be designed to have unequal class intervals.)

| 206 | 244 | 204 | 209 | 201 | 212 | 203 | 219 |
|---|---|---|---|---|---|---|---|
| 257 | 22 | 281 | 157 | 211 | 126 | 84 | 208 |
| 213 | 203 | 357 | 202 | 207 | 214 | 318 | 218 |
| 184 | 207 | 205 | 221 | 217 | 17 | 219 | 203 |
| 215 | 139 | 210 | 418 | 216 | 296 | 220 | 193 |

[There is no unique solution, but one solution is: 1–100 3; 101–200 5; 201–210 13; 211–220 11; 221–320 6; 321–420 2.]

8  The time taken in hours to the failure of 50 specimens of a metal subjected to fatigue failure tests are as shown. Form a frequency distribution, having about 8 classes and unequal class intervals, for this data.

| 28 | 22 | 23 | 20 | 12 | 24 | 37 | 28 | 21 | 25 |
|---|---|---|---|---|---|---|---|---|---|
| 21 | 14 | 30 | 23 | 27 | 13 | 23 | 7  | 26 | 19 |
| 24 | 22 | 26 | 3  | 21 | 24 | 28 | 40 | 27 | 24 |
| 20 | 25 | 23 | 26 | 47 | 21 | 29 | 26 | 22 | 33 |
| 27 | 9  | 13 | 35 | 20 | 16 | 20 | 25 | 18 | 22 |

[There is no unique solution, but one solution is: 1–10  3; 11–19  7; 20–22  12; 23–25  14; 26–28  7; 29–38  5; 39–48  2.]

9  Draw a histogram for the frequency distribution given in the solution of *Problem 7*.

[Rectangles, touching one another, having mid-points of 50.5, 150.5, 205.5, 215.5, 270.5 and 370.5. The heights of the rectangles (frequency per unit class range), are 0.03, 0.05, 1.3, 1.1, 0.06 and 0.02.]

10  Draw a histogram for the frequency distribution given in the solution to *Problem 8*.

[Rectangles, touching one another, having mid-points of 5.5, 15, 21, 24, 27, 33.5 and 43.5. The heights of the rectangles (frequency per unit class range) are 0.3, 0.78, 4, 4.67, 2.33, 0.5 and 0.2]

11  The frequency distribution for a batch of 48 resistors of similar value, measured in ohms is:

20.5–20.9  3,   21.0–21.4  10,
21.5–21.9  11,  22.0–22.4  13,
22.5–22.9  9,   23.0–23.4  2.

Form a cumulative frequency distribution for this data.

[(20.95  3), (21.45, 13), (21.95, 24), (22.45, 37), (22.95, 46), (23.45, 48)]

12  The frequency distribution for the number of hours of overtime worked by a group of craftsmen during each of 48 working weeks in a year is as shown. Form a cumulative frequency distribution for this data.

25–29  5,     30–34  4,    35–39  7,
40–44  11,    45–49  12,   50–54  8,
55–59  1.

[(29.5, 5), (34.5, 9), (39.5, 16), (44.5, 27), (49.5, 39), (54.5, 47), (59.5, 48)]

13  Draw an ogive for the data given in the solution of *Problem 11*.

14  Draw an ogive for the data given in the solution of *Problem 12*.

15  The diameter in millimetres of a reel of wire is measured in 48 places and the results are as shown.

| 2.10 | 2.29 | 2.32 | 2.21 | 2.14 | 2.22 | 2.24 | 2.05 | 2.29 | 2.18 | 2.24 | 2.16 |
|---|---|---|---|---|---|---|---|---|---|---|---|
| 2.28 | 2.18 | 2.17 | 2.20 | 2.23 | 2.13 | 2.15 | 2.22 | 2.14 | 2.27 | 2.09 | 2.21 |
| 2.26 | 2.10 | 2.21 | 2.17 | 2.28 | 2.15 | 2.11 | 2.17 | 2.22 | 2.19 | 2.12 | 2.20 |
| 2.16 | 2.25 | 2.23 | 2.11 | 2.27 | 2.34 | 2.23 | 2.07 | 2.13 | 2.26 | 2.16 | 2.12 |

(a) Form a frequency distribution of diameters having about 6 classes.

(b) Draw a histogram depicting the data.

(c) Form a cumulative frequency distribution.

(d) Draw an ogive for the data.

(a) There is no unique solution, but one solution is: 2.05–2.09 3, 2.10–2.14 10, 2.15–2.19 11, 2.20–2.24 13, 2.25–2.29 9, 2.30–2.34 2.

(b) Rectangles, touching one another, having mid-points of 2.07, 2.12, ..... and heights of 3, 10, ..... .

(c) Using the frequency distribution given in the solution to part (a) gives: 2.095 3, 2.145 13, 2.195 24, 2.245 37, 2.295 46, 2.345 48.

(d) A graph of cumulative frequency against upper class boundary having the co-ordinates given in part (c).

# 10 Measures of central tendency and dispersion

## A. MAIN POINTS CONCERNED WITH MEASURES OF CENTRAL TENDENCY AND DISPERSION

### Measures of central tendency

1. A single value, which is representative of a set of values, may be used to give an indication of the general size of the members in a set, the word 'average' often being used to indicate the single value. The statistical term used for 'average' is the arithmetic mean or just the mean. Other measures of central tendency may be used and these include the median and the modal values.

### Discrete data

2. The **arithmetic mean value** is found by adding together the values of the members of a set and dividing by the number of members in the set. Thus, the mean of the set of numbers: { 4, 5, 6, 9} is: $\frac{4+5+6+9}{4}$, i.e. 6.

   In general, the mean of the set: { $x_1, x_2, x_3, \ldots x_n$ } is

   $$\bar{x} = \frac{x_1 + x_2 + x_3 + \ldots + x_n}{n}, \text{ written as } \frac{\Sigma x}{n}$$

   where $\Sigma$ is the Greek letter 'sigma' and means 'the sum of', and $\bar{x}$, (called x-bar) is used to signify a mean value.

3. The **median value** often gives a better indication of the general size of a set containing extreme values. The set: { 7, 5, 74, 10} has a mean value of 24, which is not really representative of any of the values of the members of the set. The median value is obtained by:

   (a) **ranking** the set in ascending order of magnitude, and
   
   (b) selecting the value of the middle member for sets containing an odd number of members, or finding the value of the mean of the two middle members for sets containing an even number of members.

   For example, the set: { 7, 5, 74, 10} is ranked as { 5, 7, 10, 74}, and since it contains an even number of members (four in this case), the mean of 7 and 10 is taken, giving a median value of 8.5. Similarly, the set: { 3, 81, 15, 7, 14} is ranked as { 3, 7, 14, 15, 81} and the median value is the value of the middle member, i.e., 14.

4. The **modal value**, or just the **mode**, is the most commonly occurring value in a set. If two values occur with the same frequency, the set is 'bi-modal'. The set: { 5, 6, 8, 2, 5, 4, 6, 5, 3} has a modal value of 5, since the member having a value of 5 occurs three times. (See *Problems 1 and 2.*)

### Grouped data

5 The mean value for a set of grouped data is found by determining the sum of the (frequency × class mid-point values) and dividing by the sum of the frequencies, i.e., mean value

$$\bar{x} = \frac{f_1 x_1 + f_2 x_2 + \ldots + f_n x_n}{f_1 + f_2 + \ldots + f_n} = \frac{\Sigma(fx)}{\Sigma f}$$

where $f$ is the frequency of the class having a mid-point value of $x$, and so on, (see *Problem 3*).

6 The mean, median and modal values for grouped data may be determined from a histogram. In a histogram, frequency values are represented vertically and variable values horizontally. The mean value is given by the value of the variable corresponding to a vertical line drawn through the centroid of the histogram. The median value is obtained by selecting a variable value such that the area of the histogram to the left of a vertical line drawn through the selected variable value is equal to the area of the histogram on the right of the line. The modal value is the variable value obtained by dividing the width of the highest rectangle in the histogram in proportion to the heights of the adjacent rectangles. The method of determining the mean, median and modal values from a histogram is shown in *Problem 4*.

### Dispersion

7 The **standard deviation** of a set of data gives an indication of the amount of dispersion, or the scatter, of members of the set from the measure of central tendency. Its value is the root–mean–square value of the members of the set and for discrete data is obtained as follows:

(a) determine the measure of central tendency, usually the mean value, $\bar{x}$, (occasionally the median or modal value are specified),

(b) calculate the deviation of each member of the set from the mean, giving $(x_1 - \bar{x}), (x_2 - \bar{x}), (x_3 - \bar{x}), \ldots$

(c) determine the squares of these deviations, i.e. $(x_1 - \bar{x})^2, (x_2 - \bar{x})^2, (x_3 - \bar{x})^2, \ldots$

(d) find the sum of the squares of the deviations, that is $(x_1 - \bar{x})^2 + (x_2 - \bar{x})^2 + (x_3 - \bar{x})^2 + \ldots,$

(e) divide by the number of members in the set, $n$, giving

$$\frac{(x_1 - \bar{x})^2 + (x_2 - \bar{x})^2 + (x_3 - \bar{x})^2 + \ldots}{n}$$

(f) determine the square root of (e).

The standard deviation is indicated by $\sigma$ (the Greek letter small 'sigma') and is written mathematically as:

$$\text{standard deviation, } \sigma = \sqrt{\left\{\frac{\Sigma(x-\bar{x})^2}{n}\right\}},$$

where $x$ is a member of the set, $\bar{x}$ is the mean value of the set and $n$ is the number of members in the set. The value of standard deviation gives an indication of the distance of the members of a set from the mean value. The set: { 1, 4, 7, 10, 13} has a mean value of 7 and a standard deviation of about 4.2. The set { 5, 6, 7, 8, 9} also has a mean value of 7, but the standard deviation is about 1.4. This shows that the members of the second set are mainly much closer to the mean value than the members of the first set. The method of determining the standard deviation for a set of discrete data is shown in *Problem 5*.

8  For grouped data, standard deviation

$$\sigma = \sqrt{\left\{\frac{\Sigma\{f(x-\bar{x})^2\}}{\Sigma f}\right\}} \; ,$$

where $f$ is the class frequency value, $x$ is the class mid-point value and $\bar{x}$ is the mean value of the grouped data. The method of determining the standard deviation for a set of grouped data is shown in *Problem 6*.

9  Other measures of dispersion which are sometimes used are the quartile, decile and percentile values. The **quartile values** of a set of discrete data are obtained by selecting the values of members which divide the set into four equal parts. Thus for the set: { 2, 3, 4, 5, 5, 7, 9, 11, 13, 14, 17} there are 11 members and the values of the members dividing the set into four equal parts are 4, 7, and 13. These values are signified by $Q_1$, $Q_2$ and $Q_3$ and called the first, second and third quartile values respectively. It can be seen that the second quartile value, $Q_2$, is the value of the middle member and hence is the median value of the set.

10  For grouped data the ogive may be used to determine the quartile values. In this case, points are selected on the vertical cumulative frequency values of the ogive, such that they divide the total value of cumulative frequency into four equal parts. Horizontal lines are drawn from these values to cut the ogive. The values of the variable corresponding to these cutting points on the ogive give the quartile values (see *Problem 7*).

11  When a set contains a large number of members, the set can be split into ten parts, each containing an equal number of members. These ten parts are then called **deciles**. For sets containing a very large number of members, the set may be split into one hundred parts, each containing an equal number of members. One of these parts is called a **percentile**.

## B. WORKED PROBLEMS ON MEASURES OF CENTRAL TENDENCY AND DISPERSION

*Problem 1* Determine the mean, median and mode for the set:
{ 2, 3, 7, 5, 5, 13, 1, 7, 4, 8, 3, 4, 3}

The mean value is obtained by adding together the values of the members of the set and dividing by the number of members in the set.

Thus, mean value, $\bar{x} = \dfrac{2+3+7+5+5+13+1+7+4+8+3+4+3}{13}$

$= \dfrac{65}{13} = 5$

To obtain the median value the set is ranked, that is, placed in ascending order of magnitude, and since the set contains an odd number of members the value of the middle member is the median value.

Ranking the set gives: { 1, 2, 3, 3, 3, 4, 4, 5, 5, 7, 7, 8, 13}

The middle term is the seventh member, i.e. 4. Thus the median value is 4. The modal value is the value of the most commonly occurring member and is 3, which occurs three times, all other members only occurring once or twice.

*Problem 2* The following set of data refers to the amount of money in £'s taken by a news vendor for 6 days. Determine the mean, median and modal values of the set:

$$\{27.90,\ 34.70,\ 54.40,\ 18.92,\ 47.60,\ 39.68\}$$

Mean value $= \dfrac{27.90+34.70+54.40+18.92+47.60+39.68}{6} =$ £37.20

The ranked set is: $\{18.92,\ 27.90,\ 34.70,\ 39.68,\ 47.60,\ 54.40\}$.
Since the set has an even number of members, the mean of the middle two members is taken to give the median value,

i.e. median value $= \dfrac{34.70+39.68}{2} =$ £37.19

Since no two members have the same value, this set has **no mode**.

*Problem 3* The frequency distribution for the value of resistance in ohms of 48 resistors is as shown. Determine the mean value of resistance.
20.5–20.9   3,   21.0–21.4   10,   21.5–21.9   11
22.0–22.4   13,   22.5–22.9   9,   23.0–23.4   2

The class mid-point/frequency values are:

20.7  3,   21.2  10,   21.7  11,   22.2  13,   22.7  9 and  23.2  2.

For grouped data, the mean value is given by:

$$\bar{x} = \dfrac{\Sigma(fx)}{\Sigma f}$$

where $f$ is the class frequency and $x$ is the class mid-point value
Hence
mean value, $\bar{x}$

$= \dfrac{(3 \times 20.7)+(10 \times 21.2)+(11 \times 21.7)+(13 \times 22.2)+(9 \times 22.7)+(2 \times 23.2)}{48}$

$= \dfrac{1052.1}{48} = 21.919 \ldots$

i.e., **the mean value is 21.9 ohms**, correct to 3 significant figures.

*Problem 4* The time taken in minutes to assemble a device is measured 50 times and the results are as shown. Draw a histogram depicting this data and hence determine the mean, median and modal values of the distribution.
14.5–15.5   5,   16.5–17.5   8,   18.5–19.5   16,
20.5–21.5   12,   22.5–23.5   6,   24.6–25.5   3.

The histogram is shown in *Fig 1*. The mean value lies at the centroid of the histogram. With reference to any arbitrary axis, say YY shown at a time of 14 minutes, the position of the horizontal value of the centroid can be obtained from the relationship $AM = \Sigma(am)$, where $A$ is the area of the histogram, $M$ is the horizontal distance of the centroid from the axis YY, $a$ is the area of a rectangle of the histogram and $m$ is the distance of the centroid of the rectangle from YY (see chapter 7). The areas of the individual rectangles are shown circled on the histogram giving a total area of 100 square units. The positions, $m$, of the centroids of the individual rectangles are 1, 3, 5, . . . . . . units from YY.

Thus $100M = (10 \times 1)+(16 \times 3)+(32 \times 5)+(24 \times 7)+(12 \times 9)+(6 \times 11)$

i.e. $M = \dfrac{560}{100} = 5.6$ units from YY.

Thus the position of the mean with reference to the time scale is 14+5.6, i.e. **19.6 minutes**.

The median is the value of time corresponding to a vertical line dividing the total area of the histogram into two equal parts. The total area is 100 square units, hence the vertical line must be drawn to give 50 units of area on each side. To achieve this with reference to *Fig 1*, rectangle ABFE must be split so that

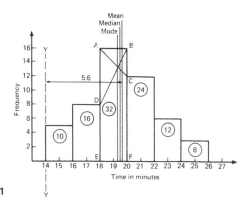

**Fig 1**

50−(10+16) units of area lie on one side and 50−(24+12+6) units of area lie on the other. This shows that the area of ABFE is split so that 24 units of area lie to the left of the line and 8 units of area lie to the right, i.e., the vertical line must pass through 19.5 minutes. Thus the median value of the distribution is **19.5 minutes**.

The mode is obtained by dividing the line AB, which is the height of the highest rectangle, proportionally to the heights of the adjacent rectangles. With reference to *Fig 1*, this is done by joining AC and BD and drawing a vertical line through the point of intersection of these two lines. This gives the mode of the distribution and is **19.3 minutes**.

*Problem 5* Determine the standard deviation from the mean of the set of numbers: {5, 6, 8, 4, 10, 3}, correct to 4 significant figures.

The arithmetic mean, $\bar{x} = \dfrac{\Sigma x}{n} = \dfrac{5+6+8+4+10+3}{6} = 6$

Standard deviation, $\sigma = \sqrt{\left\{\dfrac{\Sigma(x-\bar{x})^2}{n}\right\}}$, (see para. 7).

The $(x-\bar{x})^2$ values are: $(5-6)^2$, $(6-6)^2$, $(8-6)^2$, $(4-6)^2$, $(10-6)^2$ and $(3-6)^2$.
The sum of the $(x-\bar{x})^2$ values, i.e., $\Sigma(x-\bar{x})^2 = 1+0+4+4+16+9$
i.e. $\Sigma(x-\bar{x})^2 = 34$
and $\dfrac{\Sigma(x-\bar{x})^2}{n} = \dfrac{34}{6} = 5.\dot{6}$, since there are 6 members in the set.

Hence, standard deviation, $\sigma = \sqrt{\left\{\dfrac{\Sigma(x-\bar{x})^2}{n}\right\}}$

$\qquad = \sqrt{5.\dot{6}} = \mathbf{2.380}$, correct to 4 significant figures.

---

*Problem 6* The frequency distribution for the values of resistance in ohms of 48 resistors is as shown. Calculate the standard deviation from the mean of the resistors, correct to 3 significant figures.
20.5–20.9   3,   21.0–21.4   10,   21.5–21.9   11,
22.0–22.4   13,   22.5–22.9   9,   23.0–23.4   2.

The standard deviation for grouped data is given by:

$\sigma = \sqrt{\left\{\dfrac{\Sigma\{f(x-\bar{x})^2\}}{\Sigma f}\right\}}$, (see para. 8).

From *Problem 3*, the distribution mean value, $\bar{x} = 21.92$, correct to 4 significant figures.

The 'x-values' are the class mid-point values, i.e., 20.7, 21.2, 21.7, ........
Thus the $(x-\bar{x})^2$ values are $(20.7-21.92)^2$, $(21.2-21.92)^2$, $(21.7-21.92)^2$, .........., and the $f(x-\bar{x})^2$ values are $3(20.7-21.92)^2$, $10(21.2-21.92)^2$, $11(21.7-21.92)^2$, ..........
The $\Sigma f(x-\bar{x})^2$ values are $4.4652+5.1840+0.5324+1.0192+5.4756+3.2768$,
i.e. 19.9532.

$\dfrac{\Sigma\{f(x-\bar{x})^2\}}{\Sigma f} = \dfrac{19.9532}{48} = 0.415\,69$,

and the standard deviation, $\sigma = \sqrt{\left\{\dfrac{\Sigma\{f(x-\bar{x})^2\}}{\Sigma f}\right\}}$

$\qquad = \sqrt{0.415\,69} = \mathbf{0.645}$, correct to 3 significant figures.

---

*Problem 7* The frequency distribution given below refers to the overtime worked by a group of craftsmen during each of 48 working weeks in a year.
25–29   5,   30–34   4,   35–39   7,   40–44   11,
45–49   12,   50–54   8,   55–59   1.
Draw an ogive for this data and hence determine the quartile values.

The cumulative frequency distribution (i.e., upper class boundary/cumulative frequency values) is:

29.5  5,  34.5  9,  39.5  16,  44.5  27,  49.5  39,  54.5  47,  59.5  48.

The ogive is formed by plotting these values on a graph, as shown in *Fig 2*. The total frequency is divided into four equal parts, each having a range of 48/4, i.e., twelve. This gives cumulative frequency values of 0 to 12 corresponding to the first quartile, 12 to 24 corresponding to the second quartile, 24 to 36 corresponding to the third quartile and 36 to 48 corresponding to the fourth quartile of the distribution, i.e., the distribution is divided into four equal parts. The quartile values are those of the variable corresponding to cumulative frequency values of 12, 24 and 36, marked $Q_1$, $Q_2$ and $Q_3$ in *Fig 2*. These values, correct to the nearest hour, are **37 hours, 43 hours and 48 hours** respectively. The $Q_2$ value is also equal to the median value of the distribution. One measure of the dispersion of a distribution is called the **semi-interquartile range** and is given by $(Q_3 - Q_1)/2$, and is $(48-37)/2$ in this case, i.e. $5\frac{1}{2}$ hours.

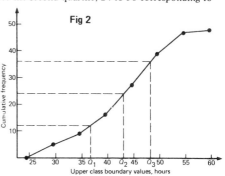

Fig 2

## C. FURTHER PROBLEMS ON MEASURES OF CENTRAL TENDENCY AND DISPERSION

In *Problems 1 to 4*, determine the mean, median and modal values for the sets given.

1  { 3, 8, 10, 7, 5, 14, 2, 9, 8}
[mean $7\frac{1}{3}$; median 8; mode 8]

2  { 26, 31, 21, 29, 32, 26, 25, 28}
[mean $27\frac{1}{4}$; median 27; mode 26]

3  { 4.72, 4.71, 4.74, 4.73, 4.72, 4.71, 4.73, 4.72}
[mean 4.7225; median 4.72; mode 4.72]

4  { 73.8, 126.4, 40.7, 141.7, 28.5, 237.4, 157.9}
[mean 115.2; median 126.4; no mode]

5  The frequency distribution given below refers to the heights in centimetres of 100 people. Determine the mean value of the distribution, correct to the nearest millimetre.

150–156  5,  157–163  18,  164–170  20
171–177  27,  178–184  22,  185–191  8.

[171.7 cm]

6  Determine the mean value of the grouped data given below, correct to 3 significant figures. The data refers to the mass in kilograms of 70 components.

0.30–0.31  4,  0.32–0.33  10,  0.34–0.35  14,
0.36–0.37  17,  0.38–0.39  12,  0.40–0.41  8,
0.42–0.43  2,  0.44–0.45  2,  0.46–0.47  1.

[0.365 kg]

7  The gain of 90 similar transistors is measured and the results are as shown.
   83.5–85.5   6,   86.5–88.5  39,   89.5–91.5  27,
   92.5–94.5  15,   95.5–97.5   3.
   By drawing a histogram of this frequency distribution, determine the mean,
   median and modal values of the distribution.
   [mean 89.5; median 89; mode 88.2]

8  The diameters, in centimetres, of 60 holes bored in engine castings are measured
   and the results are as shown. Draw a histogram depicting these results and hence
   determine the mean, median and modal values of the distribution.
   2.011–2.014   7,   2.016–2.019  16,   2.021–2.024  23,
   2.026–2.029   9,   2.031–2.034   5.
   [mean 2.021 58 cm; median 2.021 52 cm; mode 2.021 67 cm]

9  Determine the standard deviation from the mean of the set of numbers:
   { 35, 22, 25, 23, 28, 33, 30} correct to 3 significant figures.

   [4.60]

10 The values of capacitances, in microfarads, of ten capacitors selected at random
   from a large batch of similar capacitors are:
   34.3, 25.0, 30.4, 34.6, 29.6, 28.7, 33.4, 32.7, 29.0 and 31.3.
   Determine the standard deviation from the mean for these capacitors, correct to
   3 significant figures.

   [2.03 $\mu$F]

11 The tensile strength in megapascals for 15 samples of tin were determined and
   found to be:
   34.61, 34.57, 34.40, 34.63, 34.63, 34.51, 34.49, 34.61, 34.52, 34.55,
   34.58, 34.53, 34.44, 34.48 and 34.40.
   Calculate the mean and standard deviation from the mean for these 15 values,
   correct to 4 significant figures.
   [mean 34.53 MPa; standard deviation 0.074 74 MPa]

12 Determine the standard deviation from the mean, correct to 4 significant figures,
   for the heights of the 100 people given in *Problem 5*.

   [9.394 cm]

13 For the frequency distribution of the masses of 70 components given in *Problem 6*,
   Calculate the standard deviation from the mean, correct to 3 significant figures.
   [0.0346 kg]

14 Calculate the standard deviation from the mean for the data given in *Problem 7*,
   correct to 3 decimal places.

   [2.828]

15 Determine the quartile values and semi-interquartile range for the frequency
   distribution given in *Problem 5*.
   [$Q_1$ = 164.5 cm; $Q_2$ = 172.5 cm; $Q_3$ = 179 cm; 7.25 cm]

# 11 An introduction to normal distribution curves

### A. MAIN POINTS CONCERNED WITH AN INTRODUCTION TO NORMAL DISTRIBUTION CURVES

1. When data is obtained, it can frequently be considered to be a sample, (i.e., a few members), drawn at random from a large population, (i.e. a set having many members). If the sample number is large it is theoretically possible to choose class intervals which are very small, but which still have a number of members falling within each class. A frequency polygon of this data then has a large number of small line segments, and approximates to a continuous curve. Such a curve is called a **frequency** or a **distribution curve**.
2. Most distribution curves fall largely into three categories:
   (i) those having a positive skew (see *Fig 1(a)*),
   (ii) those having a negative skew (see *Fig 1(b)*), and
   (iii) those which are symmetrical (see *Fig 1(c)*).
3. An extremely important symmetrical distribution curve is called the **normal curve**, which is similar in shape to the distribution curve shown in *Fig 1(c)*. This curve can be described by a mathematical equation and is the basis of much of the work done in more advanced statistics. Many natural occurrences such as the heights or weights of a group of people, the size of components produced by a machine and the life length of certain components, approximate to a normal distribution.
4. The standard deviation of a set of data which approximates to a normal curve can be determined using the techniques introduced in chapter 10. When a normal

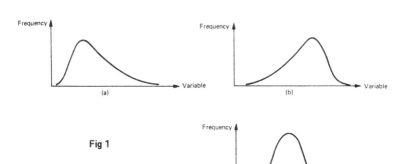

Fig 1

curve is drawn, a relationship exists between the area under parts of the curve and the value of the standard deviation of the data. Since a normal curve is symmetrical, the mean value of the data corresponds to the maximum value of the normal distribution curve.
  (i) If two vertical lines are drawn at distances corresponding to one standard deviation on either side of the mean value, then for **all** normal distribution curves, the area enclosed by the normal curve and the vertical lines at the mean value ± 1 standard deviation is approximatly two thirds or $66\frac{2}{3}\%$ of the total area beneath the curve.
  (ii) For **all** normal curves the area enclosed by the curve and vertical lines drawn at distances corresponding to the mean value ± 2 standard deviations is approximately $\frac{19}{20}$ ths or 95% of the total area beneath the curve.
  (iii) For **all** normal curves, the area enclosed by the curve and the vertical lines drawn at distances corresponding to the mean value ± 3 standard deviations is approximately $\frac{399}{400}$ ths or $99\frac{3}{4}\%$ of the total area beneath the curve.
5 The area beneath a normal curve is directly proportional to frequency and it follows that for a population which approximates to a normal distribution:
  (i) the values of $\frac{2}{3}$ rds of the members lie between the mean value ± 1 standard deviation,
  (ii) the values of $\frac{19}{20}$ ths of the members lie between the mean value ± 2 standard deviations, and
  (iii) the values of $\frac{399}{400}$ ths of the members lie between the mean value ± 3 standard deviations.
6 It should never be assumed that because data is continuous it automatically follows that it is normally distributed. One of the ways of determining whether data is normally distributed is by using **normal probability paper**, often just called **probability paper**. This is special graph paper which has linear markings on one axis and probability values from 0.01 to 99.99 on the other axis (see *Fig 4*). The divisions on the probability axis are such that a straight line graph results for normally distributed data, when percentage cumulative frequency values are plotted against upper class boundary values. The method used to test the normality of a distribution is shown in *Problems 5 and 6*.

## B. WORKED PROBLEMS ON AN INTRODUCTION TO NORMAL DISTRIBUTION CURVES

*Problem 1* The mean height of 500 people is 170 cm and the standard deviation is 9 cm. Assuming a normal distribution, determine the number of people likely to have heights of between 152 cm and 197 cm.

The mean value, $\bar{x}$ is 170 cm. Since the standard deviation, $\sigma$, is 9 cm, the mean value plus one standard deviation, written as $(\bar{x}+1\sigma)$, is 170+9, i.e. 179 cm. Similarly, $(\bar{x}+2\sigma)$ is 170+2 × 9 = 188 cm and $(\bar{x}+3\sigma)$ is 170+3 × 9 = 197 cm. Also, $(\bar{x}-1\sigma)$ is 170−9, i.e. 161 cm and $(\bar{x}-2\sigma)$ is 170−2 × 9, i.e., 152 cm. Thus the range of heights required is from $(\bar{x}-2\sigma)$ to $(\bar{x}+3\sigma)$. From para. 4, $\frac{19}{20}$ ths of the area lies between $(\bar{x}\pm2\sigma)$. Because the normal curve is symmetrical, it follows that half of $\frac{19}{20}$ ths, i.e. $\frac{19}{40}$ ths of the area lies between $\bar{x}$ and $-2\sigma$. Similarly,

$\frac{399}{400}$ths of the area lies between $(\bar{x} \pm 3\sigma)$, so half of $\frac{399}{400}$ths, i.e. $\frac{399}{800}$ths of the area lies between $\bar{x}$ and $+3\sigma$. Hence, the total area between $(\bar{x}-2\sigma)$ and $(\bar{x}+3\sigma)$ is $(\frac{19}{40} + \frac{399}{800})$ths of the total area, i.e., 0.973 75 of the total area. However, from para. 5, the area is proportional to the frequency, hence, 0.973 75 of 500 people, that is **487 people** are likely to have heights of between 152 cm and 197 cm. (In statistics, it is not usual to give results of values determined accurately, since in most instances, it is a probability which is being determined. Thus the answer to this problem is expressed correct to the nearest whole number.)

*Problem 2* For the group of people given in *Problem 1*, determine the number of people likely to have heights of less than 161 cm.

A height of 161 cm corresponds to 170−9 cm, that is, mean minus one standard deviation, $(\bar{x}-1\sigma)$. Since the normal curve is symmetrical, the area to the left of the mean value is half of the total area, (see *Fig 2(a)*). The area between $(\bar{x} \pm 1\sigma)$ is $\frac{2}{3}$rds of the total area (see para. 4). Hence the area between $\bar{x}$ and $-1\sigma$ is half of $\frac{2}{3}$rds, i.e. $\frac{1}{3}$ rd of the total area, (see *Fig 2(b)*). An area of less than $-1\sigma$, is therefore $\frac{1}{2} - \frac{1}{3}$, i.e. $\frac{1}{6}$th of the total area (see *Fig 2(c)*). However, areas are proportional to frequency (see para. 5), hence $\frac{1}{6}$th of 500, i.e., **83 people**, are likely to have heights of less than 161 cm.

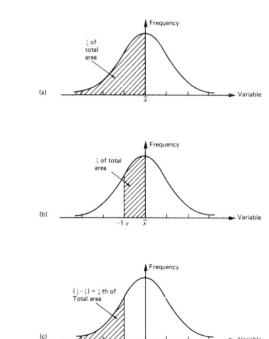

**Fig 2**

*Problem 3* For the group of people given in *Problem 1*, determine the number of people likely to have heights of more than 188 cm.

A height of 188 cm corresponds to 170+18, that is, the mean plus two standard deviations, ($\bar{x}+2\sigma$). The area to the right of the mean value is half of the total area under the normal curve (see *Fig. 3(a)*). The area between ($\bar{x} \pm 2\sigma$) is $\frac{19}{20}$ths of the total area (see para. 4). Hence the area between $\bar{x}$ and $+2\sigma$ is half of $\frac{19}{20}$ths, i.e. $\frac{19}{40}$ths of the total area (see *Fig 3(b)*). An area to the right of the $+2\sigma$ line is therefore $\frac{1}{2} - \frac{19}{40}$, i.e. 0.025 of the total area, (see *Fig 3(c)*). Since areas are proportional to frequencies, it is likely that 0.025 of 500 people, i.e. **13 people**, will have heights of more than 188 cm.

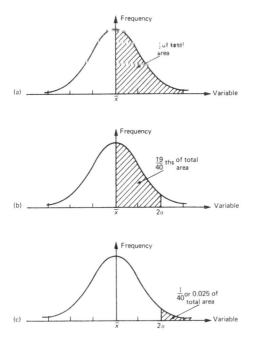

**Fig 3**

*Problem 4* A batch of lemonade bottles have an average content of 753 millilitres and a standard deviation of 1.8 millilitres. If the batch contains 1500 bottles and the contents of the bottles follow a normal distribution, determine:
(a) the number of bottles likely to contain less than 749.4 ml,
(b) the number of bottles likely to contain between 751.2 and 756.6 ml, and
(c) the number of bottles likely to contain more than 758.4 ml.

(a) Since the mean value, $\bar{x}$ is 753 and the standard deviation $\sigma$ is 1.8, 749.4 corresponds to $(\bar{x} - 2\sigma)$. The area between the vertical line drawn through $-2\sigma$ and the vertical line drawn through the mean value is half of $\frac{19}{20}$ths (see *Problem 2*), i.e. $\frac{19}{40}$ths of the area to the left of the mean value. Since the normal curve is symmetrical, the total area to the left of the mean value is half the total area, i.e. $\frac{20}{40}$ths. Thus the area under the curve to the left of the $-2\sigma$ line is $(\frac{20}{40} - \frac{19}{40})$ths, i.e. $\frac{1}{40}$th or 0.025 of the total area. Area is proportional to frequency, thus the number of bottles having less than 749.4 ml is likely to be 0.025 of 1500, that is, **38 bottles**.

(b) 751.2 corresponds to $(\bar{x} - 1\sigma)$ and the area between the mean and $-1\sigma$ for a normal curve is half of $\frac{2}{3}$rds, i.e. $\frac{1}{3}$rd of the total area. 756.6 corresponds to $(\bar{x} + 2\sigma)$ and the area between the mean and $+2\sigma$ is half of $\frac{19}{20}$ths, i.e., $\frac{19}{40}$ths of the total area. Thus, the area between $-1\sigma$ and $+2\sigma$ is $(\frac{1}{3} + \frac{19}{40})$ ths, i.e., $0.8083\dot{}$ of the total area. Thus, the number of bottles likely to have contents of between 751.2 and 756.5 ml is 0.8083 of 1500, i.e., **1213 bottles**.

(c) 758.4 ml corresponds to $(\bar{x} + 3\sigma)$ and the area between the mean and $+3\sigma$ for a normal curve is half of $\frac{399}{400}$ths, i.e., $\frac{399}{800}$ths or 0.498 75 of the total area. Thus the area greater than $+3\sigma$ is 0.500 00−0.498 75, i.e., 0.001 25 of the total area. Thus the number of bottles likely to have contents of more than 758.4 ml is 0.001 25 × 1500, that is, **2 bottles**.

---

*Problem 5* Use normal probability paper to determine whether the data given below is approximately normally distributed.

| Class mid-point value | 55 | 65 | 75 | 85 | 95 | 105 | 115 | 125 | 135 | 145 |
|---|---|---|---|---|---|---|---|---|---|---|
| Frequency | 16 | 32 | 48 | 68 | 72 | 64 | 44 | 32 | 16 | 8 |

---

To test the normality of a distribution the upper class boundary/percentage cumulative frequency values are plotted on probability paper. The upper class boundary values are: 60, 70, 80, . . . . . . . ., 140, 150. The cumulative frequency values are: 16, 48, 96, . . . . . . . ., 392, 400. The percentage cumulative frequency value for a cumulative frequency of 16 is obtained from:

$$\frac{16}{400} \times 100\% = 4\%$$

Calculating each of the percentage cumulative frequency values gives: 4, 12, 24, . . . . . . . ., 98, 100. The co-ordinates of upper class boundary/cumulative frequency values are plotted as shown in *Fig 4*. (When plotting these values, it will always be found that the last co-ordinates cannot be plotted, since the maximum value on the probability scale is 99.99.) Since all the co-ordinates except one lie approximately on a straight line, this indicates that the data given is approximately normally distributed.

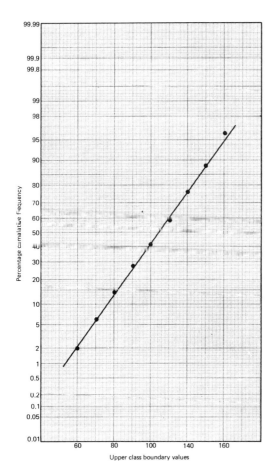

**Fig 4**

*Problem 6* The masses of 100 copper ingots are formed into a frequency distribution and the distribution is as follows:

| Class mid-point value (kg) | 29.5 | 30.5 | 31.5 | 32.5 | 33.5 | 34.5 | 35.5 | 36.5 | 37.5 | 38.5 |
|---|---|---|---|---|---|---|---|---|---|---|
| Frequency | 3 | 6 | 8 | 19 | 28 | 19 | 6 | 5 | 4 | 2 |

Use probability paper to determine whether this frequency distribution is approximately normal.

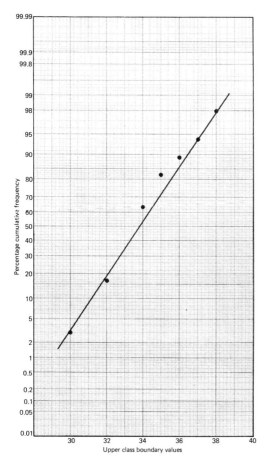

**Fig 5**

The upper class boundary values are 30, 31, 32, 33, . . . . . . . ., 38, 39. The corresponding cumulative frequency values are 3, 9, 17, 36, . . . . . . . . 98, 100. Since the total frequency is 100, the cumulative frequency values are also the percentage cumulative frequency values. The values of percentage cumulative frequencies and upper class boundaries are plotted as shown in *Fig 5*. It can be seen in *Fig 5* that a group of points at upper class boundary values of 34, 35, and 36 are displaced from a straight line drawn through the remaining co-ordinates. This shows that the distribution is not normally distributed.

## C. FURTHER PROBLEMS ON AN INTRODUCTION TO NORMAL DISTRIBUTION CURVES

1. A transistor is classed as defective if its gain is less than 69. In a batch of 350 transistors, the mean gain is 75 and the standard deviation is 3. Assuming the value of gain is normally distributed, determine how many are likely to be rejected.

   [9]

2. The masses of 800 people are normally distributed, having a mean value of 65 kg and a standard deviation of 5.5 kg. How many people are likely to have masses of less than 48.5 kg?

   [1]

3. 500 tins of paint have a mean content of 1010 ml, the standard deviation of the contents is 4 ml, and the volume of the contents is normally distributed. Calculate the number of tins which are likely to have a mass of less than (a) 1006 ml; (b) 1002 ml; and (c) 998 ml.

   [(a) 83; (b) 13; (c) 1]

4. For the batch of transistors in *Problem 1*, if those having a high gain of more than 81 are rejected, determine the number likely to be rejected due to high gain.

   [9]

5. For the 800 people in *Problem 2*, how many are likely to have masses of more than 76 kg.

   [20]

6. The diameter of holes produced by a drilling machine bit is 4.05 mm and the standard deviation of the diameters is 0.0025 mm. For 20 holes drilled using this machine, how many are likely to have diameters of between 4.045 and 4.0525 mm, assuming the diameters are normally distributed?

   [16]

7. The intelligence quotients of 400 children have a mean value of 100 and a standard deviation of 15. Assuming I.Q.'s are normally distributed, determine the number of children likely to have I.Q.'s of between: (a) 70 and 100; (b) 100 and 130; and (c) 70 and 115.

   [(a) 190; (b) 190; (c) 323]

8. The mean mass of tablets produced by a manufacturer is 5 g and the standard deviation of the masses if 0.04 g. In a bottle containing 100 tablets, determine how many are likely to have (a) a mass of less than 4.92 g; (b) a mass of between 4.88 g and 5.04 g; and (c) a mass of more than 5.08 g. Assume the masses of the tablets are normally distributed.

   [(a) 3; (b) 83; (c) 3]

9. A frequency distribution of 100 measurements is as shown:

   | Measurement: | 16.8 | 17.0 | 17.2 | 17.4 | 17.6 | 17.8 | 18.0 |
   |---|---|---|---|---|---|---|---|
   | Frequency: | 3 | 8 | 16 | 24 | 24 | 17 | 8 |

   Use normal probability paper to determine whether this distribution approximates to a normal distribution. [Yes]

10. Use probability paper to assess whether the frequency distribution given below for the diameters of 60 bolts is approximately normally distributed.

    3.93–3.94  1,  3.95–3.96  5,  3.97–3.98  9,  3.99–4.00  17,
    4.01–4.02  15,  4.03–4.04  7,  4.05–4.06  4,  4.07–4.08  2.

    [No]

# 12 Probability

## A. MAIN POINTS CONCERNED WITH PROBABILITY

1 The **probability** of something happening is the likelihood or chance of it happening. Values of probability lie between 0 and 1, where 0 represents an absolute impossibility and 1 represents an absolute certainty. The probability of an event happening usually lies somewhere between these two extreme values and is expressed either as a proper or a decimal fraction. Examples of probability are:

that a length of copper wire has zero resistance at 100°C    0

that a fair, six-sided dice will stop with a 3 upwards    $\frac{1}{6}$ or 0.1667

that a fair coin will land with a head upwards    $\frac{1}{2}$ or 0.5

that a length of copper wire has some resistance at 100°C    1.

2 If $p$ is the probability of an event happening and $q$ is the probability of the same event not happening, then the total probability is $p+q$ and is equal to unity, since it is an absolute certainty that the event either does or does not occur, i.e., $p+q = 1$.

3 The **expectation**, $E$, of an event happening is defined in general terms as the product of the probability $p$ of an event happening and the number of attempts made, $n$, i.e., $E = pn$. Thus, since the probability of obtaining a 3 upwards when rolling a fair dice is $1/6$, the expectation of getting a 3 upwards on four throws of the dice is

$\frac{1}{6} \times 4$, i.e., $\frac{2}{3}$.

Thus expectation is the average occurrence of an event.

4 A **dependent event** is one in which the probability of one event happening affects the probability of another event happening. Let 5 transistors be taken at random from a batch of 100 transistors for test purposes, and the probability of there being a defective transistor, $p_1$, be determined. At some later time, let another 5 transistors be taken at random from the 95 remaining transistors in the batch and the probability of there being a defective transistor, $p_2$, be determined. The value of $p_2$ is different from $p_1$ since the batch size has effectively altered from 100 to 95, i.e., the probability $p_2$ is dependent on probability $p_1$. Since 5 transistors are drawn, and then another 5 transistors are drawn without replacing the first 5, the second random selection is said to be **without replacement**.

5 An **independent event** is one in which the probability of an event happening does **not** affect the probability of another event happening. If 5 transistors are taken at random from a batch of transistors and the probability of a defective transistor $p_1$

is determined and the process is repeated after the original 5 have been replaced in the batch to give $p_2$, then $p_1$ is equal to $p_2$. Since the 5 transistors are replaced between draws, the second selection is said to be **with replacement**.

6 **The addition law of probability**

The addition law of probability is recognised by the word 'or' joining the probabilities. If $p_A$ is the probability of event A happening and $p_B$ is the probability of event B happening, the probability of event A or event B happening is given by $p_A + p_B$. Similarly, the probability of events A or B or C or . . . . . . . . N happening is given by

$$p_A + p_B + p_C + \ldots \ldots + p_N.$$

7 **The multiplication law of probability**

The multiplication law of probability is recognised by the word 'and' joining the probabilities. If $p_A$ is the probability of event A happening and $p_B$ is the probability of event B happening, the probability of event A and event B happening is given by $p_A \times p_B$. Similarly, the probability of events A and B and C and . . . . . N happening is given by $p_A \times p_B \times p_C \times \ldots \times p_N$.

8 The addition and multiplication laws of probability may be combined as shown below. Let $p_A$, $p_B$ and $p_C$ be the probabilities of events A, B and C respectively happening. The probabilities of events A, B and C not happening may be shown as $\bar{p}_A$, $\bar{p}_B$ and $\bar{p}_C$, (where $p_A + \bar{p}_A = 1$, see para. 2). Then, for example:
(i) the probability of events (A and B) or C happening is $(p_A \times p_B) + p_C$,
(ii) the probability of (event A or event B happening) and event C happening is $(p_A + p_B) \times p_C$,
(iii) the probability of (events A and B happening) or (event A happening and event C not happening) is $(p_A \times p_B) + (p_A \times \bar{p}_C)$,
(iv) the probability of (events A and B not happening) or (event C happening) is $(\bar{p}_A \times \bar{p}_B) + p_C$, and so on.

### B. WORKED PROBLEMS ON PROBABILITY

*Problem 1* Determine the probabilities of selecting at random (a) a man, and (b) a woman from a crowd containing 20 men and 33 women.

(a) The probability of selecting at random a man, $p$, is given by the ratio

$$\frac{\text{number of men}}{\text{number in crowd}}, \text{ i.e., } p = \frac{20}{20+33} = \frac{20}{53}.$$

(b) The probability of selecting at random a woman, $q$, is given by the ratio

$$\frac{\text{number of women}}{\text{number in crowd}}, \text{ i.e., } q = \frac{33}{20+33} = \frac{33}{53}.$$

[Check: the total probability should be equal to 1. $p = \frac{20}{53}$ and $q = \frac{33}{53}$.

The total probability, $p + q = \frac{20}{53} + \frac{33}{53} = 1$, hence no obvious error has been made.]

*Problem 2* Find the expectation of obtaining a 4 upwards with 3 throws of a fair dice.

Expectation is the average occurrence of an event and is defined as the probability times the number of attempts, (see para. 3). The probability, $p$, of obtaining a 4 upwards for one throw of the dice is $1/6$. Also, 3 attempts, $n$, are made, hence the expectation, $E$, is $pn$, i.e. $\frac{1}{6} \times 3 = \frac{1}{2}$.

*Problem 3* Calculate the probabilities of selecting at random:
(a) the winning horse in a race in which 10 horses are running,
(b) the winning horse in the first race or the winning horse in the second race if there are 10 horses in each race, and
(c) the winning horses in both the first and second races if there are 10 horses in each race.

(a) Since only one of the ten horses can win, the probability of selecting at random the winning horse is
$\frac{\text{number of winners}}{\text{number of horses}}$, i.e. $\frac{1}{10}$.

(b) The probability of selecting at random the winning horse in the first race **or** the winning horse in the second race is given by the addition law of probability, (see para. 6), since the word **or** joins the two probabilities. From (a) above, the probability of selecting the winning horse is $1/10$ for the first race and also $1/10$ for the second race. Hence, the probability of selecting the winning horse from the first **or** the second race is

$$\frac{1}{10} + \frac{1}{10} = \frac{1}{10} = \frac{1}{5} \text{ or } 0.2$$

(c) The probability of selecting the winning horse in the first race is $\frac{1}{10}$.

The probability of selecting the winning horse in the second race is $\frac{1}{10}$.

The probability of selecting the winning horses in the first **and** second race is given by the multiplication law of probability, (see para. 7),

i.e., probability $= \frac{1}{10} \times \frac{1}{10} = \frac{1}{100}$ **or 0.01**

*Problem 4* The probability of a component failing in one year due to excessive temperature is $1/20$, due to excessive vibration is $1/25$ and due to excessive humidity is $1/50$. Determine the probabilities that during a one-year period, a component:
(a) fails due to excessive temperature and excessive vibration,
(b) fails due to excessive vibration or excessive humidity and
(c) will not fail because of both excessive temperature and excessive humidity.

Let $p_A$ be the probability of failure due to excessive temperature, then $p_A = \dfrac{1}{20}$ and $\bar{p}_A = \dfrac{19}{20}$.

Let $p_B$ be the probability of failure due to excessive vibration, then $p_B = \dfrac{1}{25}$ and $\bar{p}_B = \dfrac{24}{25}$.

Let $p_C$ be the probability of failure due to excessive humidity, then $p_C = \dfrac{1}{50}$ and $\bar{p}_C = \dfrac{49}{50}$.

(a) The probability of a component failing due to excessive temperature **and** excessive vibration is given by

$p_A \times p_B$, i.e., $\dfrac{1}{20} \times \dfrac{1}{25} = \dfrac{1}{500}$ or 0.002

(b) The probability of a component failing due to excessive vibration or excessive humidity is

$p_B + p_C$, i.e. $\dfrac{1}{25} + \dfrac{1}{50} = \dfrac{3}{50}$ or 0.06

(c) The probability that a component will not fail due to excessive temperature and will not fail due to excessive humidity is

$\bar{p}_A \times \bar{p}_C$, i.e. $\dfrac{19}{20} \times \dfrac{49}{50} = \dfrac{931}{1000}$ or 0.931

---

*Problem 5* A batch of 100 capacitors contains 73 which are within the required tolerance values, 17 which are below the required tolerance values and the remainder are above the required tolerance values. Determine the probabilities that when randomly selecting a capacitor and then a second capacitor:
(a) both are within the required tolerance values when selecting with replacement and (b) the first one drawn is below and the second one drawn is above the required tolerance value, when selection is without replacement.

---

(a) The probability of selecting a capacitor within the required tolerance values is 73/100. The first capacitor drawn is now replaced and a second one is drawn from the batch of 100. The probability of this capacitor being within the required tolerance values is also 73/100. Thus, the probability of selecting a capacitor within the required tolerance values for both the first **and** the second draw is

$\dfrac{73}{100} \times \dfrac{73}{100} = \dfrac{5329}{10\,000}$ or 0.5329

(b) The probability of obtaining a capacitor below the required tolerance values on the first draw is 17/100. There are now only 99 capacitors left in the batch, since the first capacitor is not replaced. The probability of drawing a capacitor above the required tolerance values on the second draw is 10/99, since there are (100−73−17), i.e. 10 capacitors above the required tolerance value. Thus, the probability of randomly selecting a capacitor below the required tolerance values and followed by randomly selecting a capacitor above the tolerance values is

$$\frac{17}{100} \times \frac{10}{99} = \frac{170}{9900} = \frac{17}{990} = 0.0172$$

*Problem 6* A batch of 40 components contains 5 which are defective. If a component is drawn at random from the batch and tested and then a second component is drawn, determine the probability that neither of the components is defective.

**With replacement:**
The probability that the component selected on the first draw is satisfactory is 35/40, i.e. 7/8. The component is now replaced and a second draw is made. The probability that this component is also satisfactory is 7/8. Hence, the probability that both the first component drawn and the second component drawn are satisfactory is

$$\frac{7}{8} \times \frac{7}{8} = \frac{49}{64} \text{ or } 0.7656$$

**Without replacement:**
The probability that the first component drawn is satisfactory is 7/8. There are now only 34 satisfactory components left in the batch and the batch number is 39. Hence, the probability of drawing a satisfactory component on the second draw is 34/39. Thus the probability that the first component drawn and the second component drawn are satisfactory is

$$\frac{7}{8} \times \frac{34}{39} = \frac{238}{312} \text{ or } 0.7628$$

*Problem 7* A batch of 40 components contains 5 which are defective. If a component is drawn at random from the batch and tested and then a second component is drawn at random, calculate the probability of having one defective component, both with and without replacement.

The probability of having one defective component can be achieved in two ways. If $p$ is the probability of drawing a defective component and $q$ is the probability of drawing a satisfactory component, then the probability of having one defective component is given by drawing a satisfactory component and then a defective component or by drawing a defective component and then a satisfactory one, i.e., by $q \times p + p \times q$.

**With replacement:**
$p = \frac{5}{40} = \frac{1}{8}$ and $q = \frac{35}{40} = \frac{7}{8}$.

Hence, probability of having a defective component is $\frac{1}{8} \times \frac{7}{8} + \frac{7}{8} \times \frac{1}{8}$,

i.e. $\frac{7}{64} + \frac{7}{64} = \frac{14}{64} = \frac{7}{32}$ or 0.2188

**Without replacement:**

$p_1 = \frac{1}{8}$ and $q_1 = \frac{7}{8}$ on the first of the two draws. The batch number is now 39 for the second draw, thus $p_2 = \frac{5}{39}$ and $q_2 = \frac{35}{39}$

$$p_1 q_2 + q_1 p_2 = \frac{1}{8} \times \frac{35}{39} + \frac{7}{8} \times \frac{5}{39} = \frac{35+35}{312} = \frac{70}{312} \text{ or } 0.2244$$

*Problem 8* A box contains seventy-four 2.2 kΩ resistors, eighty-six 3.3 kΩ resistors and forty 4.7 kΩ resistors. Three resistors are drawn at random from the box without replacement. Determine the probability that all three are 3.3 kΩ resistors.

Assume, for clarity of explanation, that a resistor is drawn at random, then a second, then a third, (although this assumption does not affect the results obtained). The total number of resistors is 74+86+40, i.e., 200.

The probability of randomly selecting a 3.3 kΩ resistor on the first draw is 86/200. There are now 85, 3.3 kΩ resistors in a batch of 199. The probability of randomly selecting a 3.3 kΩ resistor on the second draw is 85/199. There are now 84, 3.3 kΩ resistors in a batch of 198. The probability of randomly selecting a 3.3 kΩ resistor on the third draw is 84/198. Hence the probability of selecting 3.3 kΩ resistors on the first draw and the second draw and the third draw is

$$\frac{86}{200} \times \frac{85}{199} \times \frac{84}{198} = \frac{307\,020}{3\,940\,200} = \frac{5\,117}{65\,670} \text{ or } 0.078$$

*Problem 9* For the box of resistors given in *Problem 8* above, determine the probability that there are no, 4.7 kΩ resistors drawn, when three resistors are drawn at random from the box without replacement.

The probability of not drawing a 4.7 kΩ resistor on the first draw is $1 - (40/200)$, i.e. 160/200. There are now 199 resistors in the batch of which 159 are not 4.7 kΩ resistors. Hence, the probability of not drawing a 4.7 kΩ resistor on the second draw is 159/199. Similarly, the probability of not drawing a 4.7 kΩ resistor on the third draw is $\frac{158}{198}$. Hence the probability of not drawing a 4.7 kΩ resistor on the first and second and third draws is

$$\frac{160}{200} \times \frac{159}{199} \times \frac{158}{198} = \frac{100\,488}{197\,010} = \frac{50\,244}{98\,505} \text{ or } 0.510$$

*Problem 10* For the box of resistors in *Problem 8* above, find the probability that there are two, 2.2 kΩ resistors and either a 3.3 kΩ or a 4.7 kΩ resistor when three are drawn at random, without replacement.

Two 2.2 kΩ resistors and one, 3.3 kΩ resistor can be obtained in any of the following ways:

| 1st draw | 2nd draw | 3rd draw |
|---|---|---|
| 2.2 | 2.2 | 3.3 |
| 2.2 | 3.3 | 2.2 |
| 3.3 | 2.2 | 2.2 |

Two, 2.2 kΩ resistors and one, 4.7 kΩ resistor can also be obtained in any of the following ways:

| 1st draw | 2nd draw | 3rd draw |
|---|---|---|
| 2.2 | 2.2 | 4.7 |
| 2.2 | 4.7 | 2.2 |
| 4.7 | 2.2 | 2.2 |

Thus there are six possible ways of achieving the combinations specified. If A represents a 2.2 kΩ resistor, B a 3.3 kΩ resistor and C a 4.7 kΩ resistor, then the combinations and their probabilities are as shown:

| DRAW | | | PROBABILITY |
|---|---|---|---|
| First | Second | Third | |
| A | A | B | $\frac{74}{200} \times \frac{73}{199} \times \frac{86}{198} = 0.0590$ |
| A | B | A | $\frac{74}{200} \times \frac{86}{199} \times \frac{73}{198} = 0.0590$ |
| B | A | A | $\frac{86}{200} \times \frac{74}{199} \times \frac{73}{198} = 0.0590$ |
| A | A | C | $\frac{74}{200} \times \frac{73}{199} \times \frac{40}{198} = 0.0274$ |
| A | C | A | $\frac{74}{200} \times \frac{40}{199} \times \frac{73}{198} = 0.0274$ |
| C | A | A | $\frac{40}{200} \times \frac{74}{199} \times \frac{73}{198} = 0.0274$ |

The probability of having the first combination or the second, or the third and so on is given by the sum of the probabilities, i.e., by $3 \times 0.0590 + 3 \times 0.0274$, that is, **0.2592**

## C. FURTHER PROBLEMS ON PROBABILITY

1. In a batch of 45 lamps there are 10 faulty lamps. If one lamp is drawn at random, find the probability of it being (a) faulty and (b) satisfactory.

$$\left[ (a) \frac{2}{9} ; (b) \frac{7}{9} \right]$$

2. A box of fuses are all of the same shape and size and comprises 23 2-A fuses, 47 5-A fuses and 69 13-A fuses. Determine the probability of selecting at random (a) a 2-A fuse, (b) a 5-A fuse and (c) a 13-A fuse.

$$\left[ (a) \frac{23}{139} \text{ or } 0.1655; (b) \frac{47}{139} \text{ or } 0.3381; (c) \frac{69}{139} \text{ or } 0.4964 \right]$$

3. (a) Find the probability of having a 2 upwards when throwing a fair 6-sided dice.
   (b) Find the probability of having a 5 upwards when throwing a fair 6-sided dice.

(c) Determine the probability of having a 2 and then a 5 on two throws of a fair 6-sided dice.

$$\left[\text{(a)} \frac{1}{6}; \text{(b)} \frac{1}{6}; \text{(c)} \frac{1}{36}\right]$$

4. The probability of event A happening is 3/5 and the probability of event B happening is 2/3. Calculate the probabilities of (a) both A and B happening, (b) only event A happening, i.e. event A happening and event B not happening, (c) only event B happening and (d) either A, or B, or A and B happening.

$$\left[\text{(a)} \frac{2}{5}; \text{(b)} \frac{1}{5}; \text{(c)} \frac{4}{15}; \text{(d)} \frac{13}{15}\right]$$

5. When testing 1000 soldered joints, 4 failed during a vibration test and 5 failed due to having a high resistance. Determine the probability of a joint failing due to (a) vibration, (b) high resistance, (c) vibration or high resistance and (d) vibration and high resistance.

$$\left[\text{(a)} \frac{1}{250}; \text{(b)} \frac{1}{200}; \text{(c)} \frac{9}{1000}; \text{(d)} \frac{1}{50\,000}\right]$$

6. The probability that component A will operate satisfactorily for 5 years is 4/5 and that B will operate satisfactorily over the same period of time is 3/4. Find the probabilities that in a 5 year period:
(a) both components operate satisfactorily,
(b) only component A will operate satisfactorily, and
(c) only component B will operate satisfactorily.

$$\left[\text{(a)} \frac{3}{5}; \text{(b)} \frac{1}{5}; \text{(c)} \frac{3}{20}\right]$$

7. In a particular street, 80% of the houses have telephones. If two houses selected at random are visited, calculate the probabilities that (a) they both have a telephone and (b) one has a telephone but the other does not have a telephone.

$$\left[\text{(a)} \frac{16}{25}; \text{(b)} \frac{8}{25}\right]$$

8. Veroboard pins are packed in packets of 20 by a machine. In a thousand packets, 40 have less than 20 pins. Find the probability that if 2 packets are chosen at random, one will contain less than 20 pins and the other will contain 20 pins or more.

$$\left[\frac{48}{625}\right]$$

9. A batch of 1-kW fire elements contain 16 which are within a power tolerance and 4 which are not. If 3 elements are selected at random from the batch, calculate the probabilities that (a) all three are within the power tolerance and (b) two are within but one is not within the power tolerance.

$$\left[\text{(a)} \frac{28}{57}; \text{(b)} \frac{8}{19}\right]$$

10. The statistics on numbers of boys and girls of a group of families each having 3 children are analysed. Assuming equal probability for the birth of a boy or a girl, determine the percentage of the group likely to have (a) two boys and a girl, (b) at least one boy, (c) no girls and (d) at most two girls.

[(a) 12.5%; (b) 87.5%; (c) 12.5%; (d) 87.5%]

11. An amplifier is made up of three transistors, A, B and C. The probabilities of A, B or C being defective are 1/20, 1/25 and 1/50 respectively. Calculate the percentage of amplifiers produced (a) which work satisfactorily and (b) which have just one defective transistor.

[(a) 89.38%; (b) 10.25%]

12 A box contains 14 40-W lamps, 28 60-W lamps and 58 25-W lamps, all the lamps being of the same shape and size. Three lamps are drawn at random from the box, first one, then a second, then a third. Determine the probabilities of: (a) getting one 25-W, one 40-W and one 60-W lamp, with replacement, (b) getting one 25-W, one 40-W and one 60-W lamp without replacement, (c) getting either one 25-W and two 40-W or one 60-W and two 40-W lamps with replacement and (d) getting either one 25-W and two 40-W or one 60-W and two 40-W lamps without replacement.

$$\left[ \begin{array}{l} \text{(a)} \ \frac{1421}{62\,500} \ \text{or} \ 0.0227; \ \text{(b)} \ \frac{2842}{121\,275} \ \text{or} \ 0.0234; \\ \text{(c)} \ \frac{1421}{125\,000} \ \text{or} \ 0.0114; \ \text{(d)} \ \frac{11\,739}{242\,550} \ \text{or} \ 0.0484 \end{array} \right]$$

# 13 The straight line graph

## A. MAIN POINTS CONCERNED WITH STRAIGHT LINE GRAPHS

1. A **graph** is a pictorial representation of information showing how one quantity varies with another related quantity. The most common method of showing a relationship between two sets of data is to use **cartesian** or **rectangular** axes as shown in *Fig 1*.

2. The points on a graph are called **co-ordinates**. Point A in *Fig 1* has the co-ordinates (3,2), i.e. 3 units in the $x$ direction and 2 units in the $y$ direction. Similarly, point B has co-ordinates (−4, 3) and C has co-ordinates (−3, −2). The origin has co-ordinates (0, 0).

3. The horizontal distance of a point from the vertical axis is called the **abscissa** and the vertical distance from the horizontal axis is called the **ordinate**.

4. Let a relationship between two variables $x$ and $y$ be $y = 3x+2$.
   When $x = 0$, $y = 3(0)+2 = 2$. When $x = 1$, $y = 3(1)+2 = 5$,
   When $x = 2$, $y = 3(2)+2 = 8$, and so on.
   Thus co-ordinates (0,2), (1,5) and (2,8) have been produced from the equation by selecting arbitrary values of $x$, and are shown plotted in *Fig 2*. When the points are joined together a **straight-line graph** results.

5. The **gradient or slope** of a straight line is the ratio of the change in the value of $y$ to the change in the value of $x$ between any two points on the line. If, as $x$ increases (→), $y$ also increases (↑), then the gradient is positive.

   In *Fig 3(a)* the gradient of AC = $\dfrac{\text{change in } y}{\text{change in } x} = \dfrac{\text{CB}}{\text{BA}} = \dfrac{7-3}{3-1} = \dfrac{4}{2} = 2$.

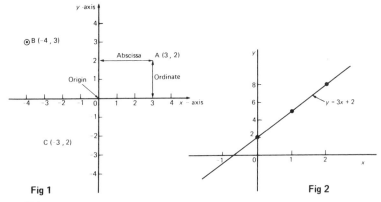

Fig 1

Fig 2

If as $x$ increases ($\rightarrow$), $y$ decreases ($\downarrow$), then the gradient is negative.

In *Fig 3(b)*, the gradient of DF = $\dfrac{\text{change in } y}{\text{change in } x} = \dfrac{\text{FE}}{\text{ED}} = \dfrac{11-2}{-3-0} = \dfrac{9}{-3} = -3$.

*Fig 3(c)* shows a straight line graph $y = 3$. Since the straight line is horizontal the gradient is zero.

**Fig 3**

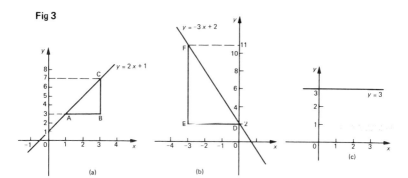

6. The value of $y$ when $x = 0$ is called the **y-axis intercept**. In *Fig 3(a)* the $y$-axis intercept is 1 and in *Fig 3(b)* is 2.

7. If the equation of a graph is of the form $y = mx+c$, where $m$ and $c$ are constants, the graph will always be a straight line, $m$ representing the gradient and $c$ the $y$-axis intercept. Thus $y = 5x+2$ represents a straight line of gradient 5 and $y$-axis intercept 2. Similarly, $y = -3x-4$ represents a straight line of gradient $-3$ and $y$-axis intercept $-4$.

8. When a set of co-ordinate values are given or are obtained experimentally and it is believed that they follow a law of the form $y = mx+c$, then if a straight line can be drawn reasonably close to most of the co-ordinate values when plotted, this verifies that a law of the form $y = mx+c$ exists. From the graph, constants $m$ (i.e. gradient) and $c$ (i.e. $y$-axis intercept) can be determined. This technique is called **determination of law**. (See *Problems 4 to 6*.)

9. (a) The process of finding co-ordinate values in between the given information is called **interpolation**.
   (b) The process of finding co-ordinate values which are outside of a given range of values is called **extrapolation**.

10. **Summary of general rules to be applied when drawing graphs**
    (i) Give the graph a title clearly explaining what is being illustrated.
    (ii) Choose scales such that the graph occupies as much space as possible on the graph paper being used.
    (iii) Choose scales so that interpolation is made as easy as possible. Usually scales such as 1 cm = 1 unit, or 1 cm = 2 units, or 1 cm = 10 units are used. Awkward scales such as 1 cm = 3 units or 1 cm = 7 units should not be used.
    (iv) The scales need not start at zero, particularly when starting at zero produces an accumulation of points within a small area of the graph paper.
    (v) The co-ordinates, or points, should be clearly marked. This may be done either by a cross, or by a dot and circle, or just by a dot (see *Fig 1*).

(vi) A statement should be made next to each axis explaining the numbers represented with their appropriate units.
(vii) Sufficient numbers should be written next to each axis without cramping.

## B. WORKED PROBLEMS ON STRAIGHT LINE GRAPHS

*Problem 1* Without plotting graphs, determine the gradient and y-axis intercept values of the following equations: (a) $y = 7x-3$; (b) $3y = -6x+2$; (c) $y-2 = 4x+9$; (d) $\frac{y}{3} = \frac{x}{3} - \frac{1}{5}$; (e) $2x+9y+1 = 0$.

(a) $y = 7x-3$ is of the form $y = mx+c$.
Hence **gradient** $m = 7$ and **y-axis intercept**, $c = -3$.

(b) Rearranging $3y = -6x+2$ gives $y = \frac{6x}{3} + \frac{2}{3}$, i.e. $y = -2x + \frac{2}{3}$ which is of the form $y = mx+c$. Hence **gradient** $m = -2$ and **y-axis intercept**, $c = \frac{2}{3}$.

(c) Rearranging $y-2 = 4x+9$ gives $y = 4x+11$.
Hence **gradient** $= 4$ and **y-axis intercept** $= 11$.

(d) Rearranging $\frac{y}{3} = \frac{x}{2} - \frac{1}{5}$ gives $y = 3\left(\frac{x}{2} - \frac{1}{5}\right) = \frac{3}{2}x - \frac{3}{5}$.
Hence **gradient** $= \frac{3}{2}$ and **y-axis intercept** $= \frac{-3}{5}$.

(e) Rearranging $2x+9y+1 = 0$ gives $9y = -2x-1$, i.e. $y = \frac{-2}{9}x - \frac{1}{9}$
Hence **gradient** $= \frac{-2}{9}$ and **y-axis intercept** $= \frac{-1}{9}$.

*Problem 2* Determine the gradient of the straight line graph passing through the co-ordinates (a) $(-2,5)$ and $(3,4)$; (b) $(-2,-3)$ and $(-1,3)$.

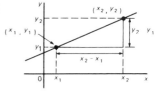

Fig 4

A straight line graph passing through co-ordinates $(x_1, y_1)$ and $(x_2, y_2)$ has a gradient given by

$m = \frac{y_2 - y_1}{x_2 - x_1}$ (see *Fig 4*.)

(a) A straight line passes through $(-2,5)$ and $(3,4)$, hence $x_1 = -2$, $y_1 = 5$, $x_2 = 3$ and $y_2 = 4$.
Hence gradient $m = \frac{y_2 - y_1}{x_2 - x_1} = \frac{4-5}{3-(-2)} = -\frac{1}{5}$

(b) A straight line passes through $(-2, -3)$ and $(-1, 3)$, hence $x_1 = -2$, $y_1 = -3$, $x_2 = -1$ and $y_2 = 3$

Hence gradient $m = \dfrac{y_2 - y_1}{x_2 - x_1} = \dfrac{3 - (-3)}{-1 - (-2)} = \dfrac{3 + 3}{-1 + 2} = \dfrac{6}{1} = 6$

*Problem 3* Plot the graphs $3x + y + 1 = 0$ and $2y - 5 = x$ on the same axes and find their point of intersection.

Rearranging $3x + y + 1 = 0$ gives $y = -3x - 1$

Rearranging $2y - 5 = x$ gives $2y = x + 5$ and $y = \dfrac{1}{2}x + 2\dfrac{1}{2}$.

Since both equations are of the form $y = mx + c$ both are straight lines.

Knowing an equation is a straight line means that only two co-ordinates need be plotted and a straight line drawn through them. A third co-ordinate is usually determined to act as a check. A table of values is produced for each equation as shown below

| $x$ | 1 | 0 | $-1$ |
|---|---|---|---|
| $-3x - 1$ | $-4$ | $-1$ | 2 |

| $x$ | 2 | 0 | $-3$ |
|---|---|---|---|
| $\dfrac{1}{2}x + 2\dfrac{1}{2}$ | $3\dfrac{1}{2}$ | $2\dfrac{1}{2}$ | 1 |

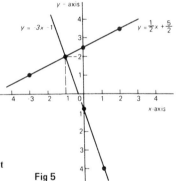

Fig 5

The graphs are plotted as shown in *Fig 5*.
**The two straight lines are seen to intersect at $(-1, 2)$.**

*Problem 4* A test on a filament lamp gave the following values of voltage $V$ for various values of resistance $R$.

| $R$ ohms | 35 | 51 | 74 | 105 | 127 |
|---|---|---|---|---|---|
| $V$ volts | 17 | 31 | 49 | 74 | 93 |

Choose suitable scales and plot a graph of $R$ representing the vertical axis and $V$ the horizontal axis. Determine (a) the gradient of the graph, (b) the $R$-axis intercept, (c) the law of the graph, (d) the value of resistance when the voltage is 40 volts, (e) the value of the voltage when the resistance is 90 ohms. (f) If the graph were to continue in the same manner, what value of resistance would be obtained at 110 V?

The co-ordinates (17,35), (31,51), (49,74), and so on are shown plotted in *Fig 6* where the best straight line is drawn through the points.

(a) Gradient of the straight line = $\dfrac{AB}{BC} = \dfrac{135-15}{100-0} = \dfrac{120}{100} = 1.2$

Note that the vertical line AB and the horizontal line BC may be constructed anywhere along the length of the straight line. However, calculations are easier if the horizontal line is carefully chosen (in this case, 100).

(b) The $R$-axis intercept is at $R = $ **15 ohms** (by extrapolation).
(c) The law of the graph is $R = 1.2V+15$, since the gradient is 1.2 and the $R$-axis intercept is 15.

From *Fig 6*, by interpolation,

(d) when the voltage is 40 V, the resistance is **63 ohms**,
(e) when the resistance is 90 ohms, the voltage is **62.5 V**, and
(f) by extrapolation, when the voltage is 110 V, the resistance is **147 ohms**.

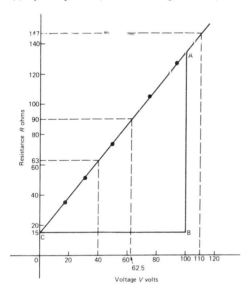

Fig 6 Graph of resistance against voltage

*Problem 5* The volume of a gas, $V$ m$^3$, was measured at various temperatures $t°$C and the results were:

| Volume $V$ m$^3$ | 26.10 | 26.75 | 27.25 | 27.90 | 28.53 | 29.10 |
|---|---|---|---|---|---|---|
| temperature $t°$C | 10 | 20 | 30 | 40 | 50 | 60 |

Prove that the volume and temperature are related by a law of the form $V = at+b$, where $a$ and $b$ are constants, and determine the law. From the graph, determine the volume at 27°C and the temperature when the volume is 29.5 m$^3$, assuming the law is true outside of the range of values given.

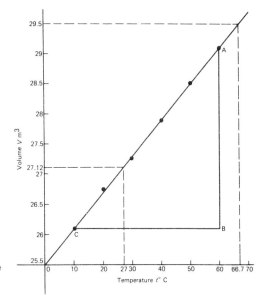

**Fig 7 Graph of volume against temperature**

Since the law relating $V$ and $t$ is $V = at+b$ (which compares with $y = mx+c$) then $V$ is plotted vertically and $t$ horizontally. If the scales are selected with each starting at zero, the co-ordinates would be bunched up at the top of the graph. Thus the co-ordinates (10,26.10), (20,26.75), and so on, are plotted as shown in *Fig 7*. The best straight line is drawn through the points. Since a straight line results this is proof that volume and temperature are related by the law $V = at+b$.

Gradient, $a = \dfrac{AB}{BC} = \dfrac{29.10-26.10}{60-10} = \dfrac{3.0}{50} = 0.06$, $V$-axis intercept $= 25.5$.

Hence the law of the graph is $V = \mathbf{0.06}t + \mathbf{25.5}$
From *Fig 7*, when the temperature is 27°C, the volume is **27.12 m³**, and when the volume is 29.5 m³, the temperature is **66.7°C**.

*Problem 6* Experimental tests to determine the breaking stress σ of rolled copper at various temperatures $t$ gave the following results.

| Stress σ N/cm² | 8.42 | 8.02 | 7.75 | 7.35 | 7.06 | 6.63 |
|---|---|---|---|---|---|---|
| Temperature $t$°C | 70 | 200 | 280 | 410 | 500 | 640 |

Show that the values obey the law $σ = at+b$, where $a$ and $b$ are constants and determine approximate values for $a$ and $b$. Use the law to determine the stress at 250°C and the temperature when the stress is 7.54 N/cm².

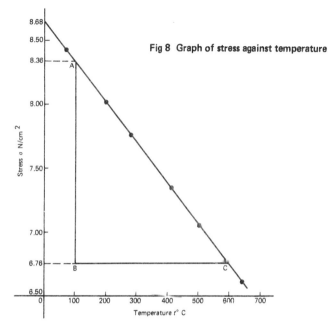

Fig 8 Graph of stress against temperature

The co-ordinates (70,8.46), (200,8.04), and so on, are plotted as shown in *Fig 8*. Since the graph is a straight line then the values obey the law $\sigma = at+b$.

Gradient of straight line, $a = \dfrac{AB}{BC} = \dfrac{8.36-6.76}{100-600} = \dfrac{1.60}{-500} = -0.0032$

Vertical axis intercept, $b = 8.68$

Hence the law of the graph is $\sigma = -0.0032t+8.68$

When the temperature is 250°C,

stress $\sigma = -0.0032(250)+8.68 = 7.88$ N/cm$^2$.

Rearranging $\sigma = -0.0032t+8.68$ gives

$0.0032t = 8.68-\sigma$ i.e. $t = \dfrac{8.68-\sigma}{0.0032}$

Hence when the stress $\sigma = 7.54$ N/cm$^2$, temperature $t = \dfrac{8.68-7.54}{0.0032} = 356.3°C$.

## C. FURTHER PROBLEMS ON STRAIGHT LINE GRAPHS

Determine the gradient and y-axis intercept for each of the equations in *Problems 1 and 2* and sketch the graphs.

1 (a) $y = 6x-3$; (b) $y = -2x+4$; (c) $y = 3x$; (d) $y = 7$

[(a) 6, $-3$; (b) $-2$, 4; (c) 3, 0; (d) 0, 7]

2 (a) $2y+1 = 4x$; (b) $2x+3y+5 = 0$; (c) $3(2y-4) = x/3$;
 (d) $5x-\dfrac{y}{2}-\dfrac{7}{3} = 0$.

$\left[\text{(a) } 2, -\dfrac{1}{2}; \text{(b) } -\dfrac{2}{3}, -1\dfrac{2}{3}; \text{(c) } \dfrac{1}{18}, 2; \text{(d) } 10, -4\dfrac{2}{3}\right]$

3. Determine the gradient of the straight line graphs passing through the co-ordinates (a) (2, 7) and (−3, 4); (b) (−4, −1) and (−5, 3); (c) $\left(\frac{1}{4}, -\frac{3}{4}\right)$ and $\left(-\frac{1}{2}, \frac{5}{8}\right)$.  $\left[\text{(a)} \frac{3}{5}; \text{(b)} -4; \text{(c)} -1\frac{5}{6}\right]$

4. State which of the following equations will produce graphs which are parallel to one another: (a) $y-4 = 2x$; (b) $4x = -(y+1)$; (c) $x = \frac{1}{2}(y+5)$; (d) $1+\frac{1}{2}y = \frac{3}{2}x$; (e) $2x = \frac{1}{2}(7-y)$
[(a) and (c); (b) and (e)]

5. Draw a graph of $y-3x+5 = 0$ over a range of $x = -3$ to $x = 4$. Hence determine (a) the value of $y$ when $x = 1.3$ and (b) the value of $x$ when $y$ is −9.2.
[(a) −1.1; (b) −1.4]

6. Plot the graphs $y = 2x+3$ and $2y = 15-2x$ on the same axes and determine their point of intersection.  $[(1\frac{1}{2}, 6)]$

7. The velocity $v$ of a body after varying time intervals $t$ was measured as follows:

| $t$ seconds | 2 | 5 | 7 | 10 | 14 | 17 |
|---|---|---|---|---|---|---|
| $v$ m/s | 15.5 | 17.3 | 18.5 | 20.3 | 22.7 | 24.5 |

Plot a graph with velocity vertical and time horizontal. Determine from the graph (a) the gradient; (b) the vertical axis intercept; (c) the equation of the graph; (d) the velocity after 12.5 s and (e) the time when the velocity is 18 m/s.
[(a) 0.6; (b) 14.3; (c) $v = 0.6t+14.3$; (d) 21.8 m/s; (e) 6.17 s]

8. In an experiment demonstrating Hooke's law, the strain in a copper wire was measured for various stresses. The results were:

| Stress (pascals) | $10.6 \times 10^6$ | $18.2 \times 10^6$ | $24.0 \times 10^6$ | $30.7 \times 10^6$ | $39.4 \times 10^6$ |
|---|---|---|---|---|---|
| Strain | 0.000 11 | 0.000 19 | 0.000 25 | 0.000 32 | 0.000 41 |

Plot a graph of stress (vertically) against strain (horizontally). Determine (a) Young's Modulus of Elasticity for copper, which is given by the gradient of the graph, (b) the value of strain at a stress of $21 \times 10^6$ Pa, (c) the value of stress when the strain is 0.000 30.
[(a) $96 \times 10^9$ Pa; (b) 0.000 22; (c) $28.8 \times 10^6$ Pa]

9. An experiment with a set of pulley blocks gave the following results:

| Effort, $E$ (Newtons) | 9.0 | 11.0 | 13.6 | 17.4 | 20.8 | 23.6 |
|---|---|---|---|---|---|---|
| Load, $L$ (Newtons) | 15 | 25 | 38 | 57 | 74 | 88 |

Plot a graph of effort (vertically) against load (horizontally) and determine (a) the gradient, (b) the vertical axis intercept, (c) the law of the graph, (d) the effort when the load is 30 N and (e) the load when the effort is 19 N.
$\left[\text{(a)} \frac{1}{5}; \text{(b)} 6; \text{(c)} E = \frac{1}{5}L+6; \text{(d)} 12 \text{ N}; \text{(e)} 65 \text{ N}\right]$

10. The mass $m$ of a steel joist varies with length $l$ as follows:

| Mass $m$ kg | 82.9 | 88.3 | 92.7 | 102.8 | 112.1 | 119.3 |
|---|---|---|---|---|---|---|
| length $l$ m | 3.24 | 3.45 | 3.62 | 4.01 | 4.37 | 4.65 |

Show that the law relating mass and length is of the form $m = al+b$ where $a$ and $b$ are constants, and determine the law.
$[m = 25.8\,l-0.67]$

11 The variation of pressure $p$ in a vessel with temperature $T$ is believed to follow a law of the form $p = aT+b$, where $a$ and $b$ are constants. Verify this law for the results given below and determine the approximate values of $a$ and $b$. Hence determine the pressures at temperatures of 285 K and 310 K and the temperature at a pressure of 250 kPa.

| pressure, $p$ kPa | 244 | 247 | 252 | 258 | 262 | 267 |
|---|---|---|---|---|---|---|
| temperature, $T$ K | 273 | 277 | 282 | 289 | 294 | 300 |

[$a = 0.85, b = 12$; 254.3 kPa, 275.5 kPa; 280 K]

# 14 Reduction of non-linear laws to linear forms

## A. MAIN POINTS CONCERNED WITH THE REDUCTION OF NON-LINEAR LAWS TO LINEAR FORM

1 Frequently, the relationship between two variables, say $x$ and $y$, is not a linear one, i.e. when $x$ is plotted against $y$ a curve results. In such cases the non-linear equation may be modified to the linear form, $y = mx+c$, so that the constants, and thus the law relating the variables can be determined. This technique is called '**determination of law**'.

2 Some examples of the reduction of equations to linear form include:

(i) $y = ax^2 + b$ compares with $Y = mX+c$, where $m = a$, $c = b$ and $X = x^2$.
Hence $y$ is plotted vertically against $x^2$ horizontally to produce a straight line graph of gradient '$a$' and $y$-axis intercept '$b$'.

(ii) $y = \dfrac{a}{x} + b$.

$y$ is plotted vertically against $\dfrac{1}{x}$ horizontally to produce a straight line graph of gradient '$a$' and $y$-axis intercept '$b$'.

(iii) $y = ax^2 + bx$

Dividing both sides by $x$ gives $\dfrac{y}{x} = ax + b$.

Comparing with $Y = mX+c$ shows that $\dfrac{y}{x}$ is plotted vertically against $x$ horizontally to produce a straight line graph of gradient '$a$' and $\dfrac{y}{x}$ axis intercept '$b$'.

(iv) $y = ax^n$

Taking logarithms to a base of 10 of both sides gives:

$\lg y = \lg (ax^n) = \lg a + \lg x^n$
i.e. $\lg y = n \lg x + \lg a$
which compares with $Y = mX+c$,

which shows that $\lg y$ is plotted vertically against $\lg x$ horizontally to produce a straight line graph of gradient $n$ and $\lg y$-axis intercept $\lg a$.

(v) $y = ab^x$

Taking logarithms to a base of 10 of both sides gives:

$\lg y = \lg (ab^x)$
i.e. $\lg y = \lg a + \lg b^x$
i.e. $\lg y = x \lg b + \lg a$
or $\lg y = (\lg b)x + \lg a$
which compares with $Y = mX+c$,

which shows that $\lg y$ is plotted vertically against $x$ horizontally to produce a straight line graph of gradient $\lg b$ and $\lg y$-axis intercept $\lg a$.

(vi) $y = ae^{bx}$

Taking logarithms to a base of $e$ of both sides gives:
$\ln y = \ln (ae^{bx})$
i.e. $\ln y = \ln a + \ln e^{bx}$
i.e. $\ln y = \ln a + bx \ln e$
i.e. $\ln y = bx + \ln a$
which compares with $Y = mX + c$,
which shows that $\ln y$ is plotted vertically against $x$ horizontally to produce a straight line graph of gradient $b$ and $\ln y$-axis intercept $\ln a$.

## B. WORKED PROBLEMS ON REDUCING NON-LINEAR LAWS TO LINEAR FORM

*Problem 1* Experimental values of $x$ and $y$, shown below, are believed to be related by the law $y = ax^2 + b$. By plotting a suitable graph verify this law and determine approximate values of $a$ and $b$.

| $x$ | 1 | 2 | 3 | 4 | 5 |
|---|---|---|---|---|---|
| $y$ | 9.8 | 15.2 | 24.2 | 36.5 | 53.0 |

If $y$ is plotted against $x$ a curve results and it is not possible to determine the values of constants $a$ and $b$ from the curve. Comparing $y = ax^2 + b$ with $Y = mX + c$ shows that $y$ is to be plotted vertically against $x^2$ horizontally. A table of values is drawn up as shown below.

| $x$ | 1 | 2 | 3 | 4 | 5 |
|---|---|---|---|---|---|
| $x^2$ | 1 | 4 | 9 | 16 | 25 |
| $y$ | 9.8 | 15.2 | 24.2 | 36.5 | 53.0 |

A graph of $y$ against $x^2$ is shown in *Fig 1*, with the best straight line drawn through the points. Since a straight line graph results, the law is verified.

From the graph, gradient, $a = \dfrac{AB}{BC} = \dfrac{53-17}{25-5} = \dfrac{36}{20} = 1.8$,

and the $y$-axis intercept, $b = 8.0$.
Hence the law of the graph is $y = 1.8x^2 + 8.0$.

*Problem 2* Values of load $L$ newtons and distance $d$ metres obtained experimentally are shown in the following table.

| Load, $L$ N | 32.3 | 29.6 | 27.0 | 23.2 | 18.3 | 12.8 | 10.0 | 6.4 |
|---|---|---|---|---|---|---|---|---|
| distance, $d$ m | 0.75 | 0.37 | 0.24 | 0.17 | 0.12 | 0.09 | 0.08 | 0.07 |

Verify that load and distance are related by a law of the form $L = \dfrac{a}{d} + b$ and determine approximate values of $a$ and $b$. Hence calculate the load when the distance is 0.20 m and the distance when the load is 20 N.

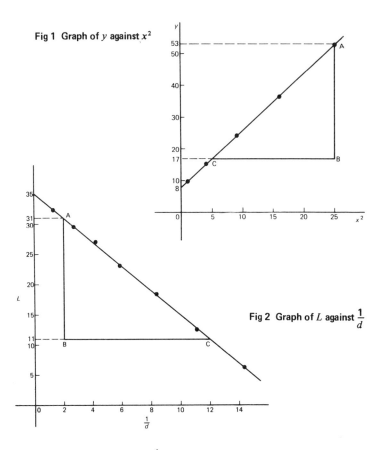

Fig 1 Graph of $y$ against $x^2$

Fig 2 Graph of $L$ against $\frac{1}{d}$

Comparing $L = \frac{a}{d} + b$, i.e. $L = a\,\frac{1}{d} + b$, with $Y = mX + c$ shows that $L$ is to be plotted vertically against $\frac{1}{d}$ horizontally. Another table of values is drawn up as shown below.

| $L$ | 32.3 | 29.6 | 27.0 | 23.2 | 18.3 | 12.8 | 10.0 | 6.4 |
|---|---|---|---|---|---|---|---|---|
| $d$ | 0.75 | 0.37 | 0.24 | 0.17 | 0.12 | 0.09 | 0.08 | 0.07 |
| $\frac{1}{d}$ | 1.33 | 2.70 | 4.17 | 5.88 | 8.33 | 11.1 | 12.5 | 14.3 |

A graph of $L$ against $\frac{1}{d}$ is shown in *Fig 2*. A straight line can be drawn through the points, which verifies that load and distance are related by a law of the form $L = \frac{a}{d} + b$.

Gradient of straight line, $a = \dfrac{\text{AB}}{\text{BC}} = \dfrac{31-11}{2-12} = \dfrac{20}{-10} = -2$

$L$-axis intercept, $b = 35$

Hence the law of the graph is $L = -\dfrac{2}{d} + 35$

When the distance $d = 0.20$ m, load $L = \dfrac{-2}{0.20} + 35 = 25.0$ N

Rearranging $L = \dfrac{-2}{d} + 35$ gives $\dfrac{2}{d} = 35 - L$ and $d = \dfrac{2}{35-L}$

Hence when load $L = 20$ N, distance, $d = \dfrac{2}{35-20} = \dfrac{2}{15} = 0.133$ m

---

*Problem 3* The solubility $s$ of potassium chlorate is shown by the following table:

| $t°C$ | 10 | 20 | 30 | 40 | 50 | 60 | 80 | 100 |
|---|---|---|---|---|---|---|---|---|
| $s$ | 4.9 | 7.6 | 11.1 | 15.4 | 20.4 | 26.4 | 40.6 | 58.0 |

The relationship between $s$ and $t$ is thought to be of the form $s = 3 + at + bt^2$. Plot a graph to test the supposition and use the graph to find approximate values of $a$ and $b$. Hence calculate the solubility of potassium chlorate at $70°C$.

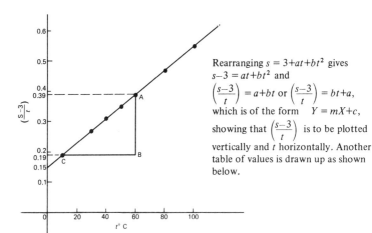

Rearranging $s = 3 + at + bt^2$ gives
$s - 3 = at + bt^2$ and
$\left(\dfrac{s-3}{t}\right) = a + bt$ or $\left(\dfrac{s-3}{t}\right) = bt + a$,
which is of the form $Y = mX + c$,
showing that $\left(\dfrac{s-3}{t}\right)$ is to be plotted vertically and $t$ horizontally. Another table of values is drawn up as shown below.

Fig 3

| $t$ | 10 | 20 | 30 | 40 | 50 | 60 | 80 | 100 |
|---|---|---|---|---|---|---|---|---|
| $s$ | 4.9 | 7.6 | 11.1 | 15.4 | 20.4 | 26.4 | 40.6 | 58.0 |
| $\left(\dfrac{s-3}{t}\right)$ | 0.19 | 0.23 | 0.27 | 0.31 | 0.35 | 0.39 | 0.47 | 0.55 |

A graph of $\left(\dfrac{s-3}{t}\right)$ against $t$ is shown plotted in *Fig 3*.

A straight line fits the points which shows that $s$ and $t$ are related by $s = 3+at+bt^2$.

Gradient of straight line, $b = \dfrac{AB}{BC} = \dfrac{0.39-0.19}{60-10} = \dfrac{0.20}{50} = 0.004$

Vertical axis intercept, $a = 0.15$
Hence the law of the graph is $s = 3+0.15t+0.004t^2$
The solubility of potassium chlorate at 70°C is given by
$s = 3+0.15(70)+0.004(70)^2 = 3+10.5+19.6 = \mathbf{33.1}$

---

*Problem 4* The current flowing in, and the power dissipated by a resistor are measured experimentally for various values and the results are as shown below.

| Current, $I$ amperes | 2.2 | 3.6 | 4.1 | 5.6 | 6.8 |
|---|---|---|---|---|---|
| Power, $P$ watts | 116 | 311 | 403 | 753 | 1110 |

Show that the law relating current and power is of the form $P = RI^n$, where $R$ and $n$ are constants, and determine the law.

---

Taking logarithms to a base of 10 of both sides of $P = RI^n$ gives:

$\lg P = \lg(RI^n) = \lg R + \lg I^n = \lg R + n \lg I$

i.e. $\lg P = n \lg I + \lg R$, which is of the form

$Y = mX + c$, showing that $\lg P$ is to be plotted vertically against $\lg I$ horizontally. A table of values for $\lg I$ and $\lg P$ is drawn up as shown below.

| $I$ | 2.2 | 3.6 | 4.1 | 5.6 | 6.8 |
|---|---|---|---|---|---|
| $\lg I$ | 0.342 | 0.556 | 0.613 | 0.748 | 0.833 |
| $P$ | 116 | 311 | 403 | 753 | 1110 |
| $\lg P$ | 2.064 | 2.493 | 2.605 | 2.877 | 3.045 |

A graph of $\lg P$ against $\lg I$ is shown in *Fig 4* and since a straight line results the law $P = RI^n$ is verified.

Gradient of straight line,
$n = \dfrac{AB}{BC} = \dfrac{2.98-2.18}{0.8-0.4}$
$= \dfrac{0.80}{0.4} = 2$

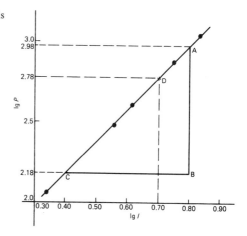

Fig 4

It is not possible to determine the vertical axis intercept on sight since the horizontal axis scale does not start at zero. Selecting any point from the graph, say point D, where lg $I$ = 0.70 and lg $P$ = 2.78, and substituting values into lg $P = n$ lg $I$+lg $R$ gives

$$2.78 = (2)(0.70)+\lg R$$
from which  lg $R$ = 2.78−1.40 = 1.38.
Hence $R$ = antilog 1.38 (= $10^{1.38}$) = 24.0
**Hence the law of the graph is $P = 24.0\ I^2$**

---

*Problem 5* The periodic time, $T$, of oscillation of a pendulum is believed to be related to its length, $l$, by a law of the form $T = kl^n$, where $k$ and $n$ are constants. Values of $T$ were measured for various lengths of the pendulum and the results are as shown below.

| Periodic time, $T$ s | 1.0 | 1.3 | 1.5 | 1.8 | 2.0 | 2.3 |
|---|---|---|---|---|---|---|
| Length, $l$ m | 0.25 | 0.42 | 0.56 | 0.81 | 1.0 | 1.32 |

Show that the law is true and determine the approximate values of $k$ and $n$. Hence find the periodic time when the length of the pendulum is 0.75 m.

---

From para 2(iv), if  $T = kl^n$ then lg $T = n$ lg $l$+lg $k$
and comparing with $\quad Y = mX + c$
shows that lg $T$ is plotted vertically against lg $l$ horizontally. A table of values for lg $T$ and lg $l$ is drawn up as shown below.

| $T$ | 1.0 | 1.3 | 1.5 |
|---|---|---|---|
| lg $T$ | 0 | 0.114 | 0.176 |
| $l$ | 0.25 | 0.42 | 0.56 |
| lg $l$ | −0.602 | −0.377 | −0.252 |

| $T$ | 1.8 | 2.0 | 2.3 |
|---|---|---|---|
| lg $T$ | 0.255 | 0.301 | 0.362 |
| $l$ | 0.81 | 1.0 | 1.32 |
| lg $l$ | −0.092 | 0 | 0.121 |

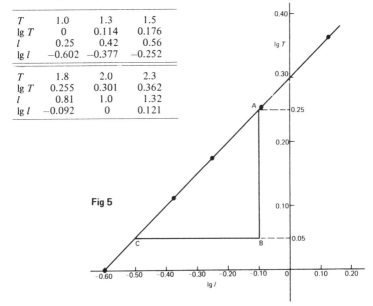

Fig 5

A graph of lg $T$ against lg $l$ is shown in *Fig 5* and the law $T = kl^n$ is true since a straight line results.

From the graph, gradient of straight line, $n = \dfrac{AB}{BC} = \dfrac{0.25 - 0.05}{-0.10 - (-0.50)} = \dfrac{0.20}{0.40} = \dfrac{1}{2}$

Vertical axis intercept, lg $k = 0.30$. Hence $k = $ antilog $0.30$ $(= 10^{0.30}) = 2.0$
**Hence the law of the graph is $T = 2.0l^{1/2}$ or $T = 2.0\sqrt{l}$**
When length $l = 0.75$ m then $T = 2.0\sqrt{(0.75)} = $ **1.73 s**

*Problem 6* Quantities $x$ and $y$ are believed to be related by a law of the form $y = ab^x$, where $a$ and $b$ are constants. Values of $x$ and corresponding values of $y$ are:

| $x$ | 0 | 0.6 | 1.2 | 1.8 | 2.4 | 3.0 |
|---|---|---|---|---|---|---|
| $y$ | 5.0 | 9.67 | 18.7 | 36.1 | 69.8 | 135.0 |

Verify the law and determine the approximate values of $a$ and $b$. Hence determine (a) the value of $y$ when $x$ is 2.1 and (b) the value of $x$ when $y$ is 100.

From para. 2(v), if $y = ab^x$ then  lg $y = ($lg $b)x + $lg $a$
and comparing with $\qquad\qquad Y = mX + c$
shows that lg $y$ is plotted vertically and $x$ horizontally. Another table is drawn up as shown below.

| $x$ | 0 | 0.6 | 1.2 |
|---|---|---|---|
| $y$ | 5.0 | 9.67 | 18.7 |
| lg $y$ | 0.70 | 0.99 | 1.27 |
| $x$ | 1.8 | 2.4 | 3.0 |
| $y$ | 36.1 | 69.8 | 135.0 |
| lg $y$ | 1.56 | 1.84 | 2.13 |

A graph of lg $y$ against $x$ is shown in *Fig 6* and since a straight line results, the law $y = ab^x$ is verified.

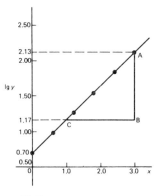

Fig 6

Gradient of straight line, lg $b = \dfrac{AB}{BC} = \dfrac{2.13 - 1.17}{3.0 - 1.0} = \dfrac{0.96}{2.0} = 0.48$

Hence $b = $ antilog $0.48$ $(= 10^{0.48}) = $ **3.0**, correct to 2 significant figures.
Vertical axis intercept, lg $a = 0.70$, from which $a = $ antilog $0.70$ $(= 10^{0.70})$
$\qquad\qquad\qquad\qquad\qquad\qquad = $ **5.0**, correct to 2 significant figures.
**Hence the law of the graph is $y = 5.0\,(3.0)^x$**

(a) When $x = 2.1$, $y = 5.0 \, (3.0)^{2.1} = \mathbf{50.2}$

(b) When $y = 100$, $100 = 5.0 \, (3.0)^x$, from which $100/5.0 = (3.0)^x$, i.e. $20 = (3.0)^x$

Taking logarithms of both sides gives $\lg 20 = \lg (3.0)^x = x \lg 3.0$

Hence $x = \dfrac{\lg 20}{\lg 3.0} = \dfrac{1.3010}{0.4771} = \mathbf{2.73}$

---

*Problem 7* The current $i$ mA flowing in a capacitor which is being discharged varies with time $t$ ms as shown below.

| $i$ mA | 203 | 61.14 | 22.49 | 6.13 | 2.49 | 0.615 |
|---|---|---|---|---|---|---|
| $t$ ms | 100 | 160 | 210 | 275 | 320 | 390 |

Show that these results are related by a law of the form $i = Ie^{t/T}$, where $I$ and $T$ are constants. Determine the approximate values of $I$ and $T$.

---

Taking Naperian logarithms of both sides of $i = Ie^{t/T}$ gives

$$\ln i = \ln (Ie^{t/T}) = \ln I + \ln e^{t/T}$$

i.e. $\ln i = \ln I + \dfrac{t}{T}$ (since $\ln e = 1$)

or $\ln i = \left(\dfrac{1}{T}\right) t + \ln I$

which compares with

$y = mx + c$, showing that $\ln i$ is plotted vertically against $t$ horizontally. (For methods of evaluating Naperian logarithms see chapter 18.) Another table of values is drawn up as shown below.

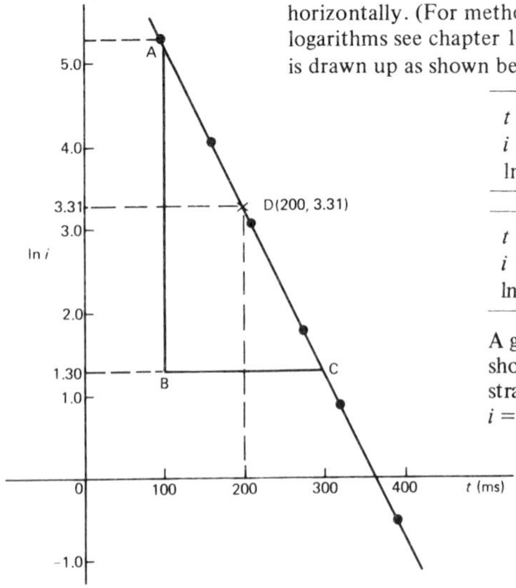

| $t$ | 100 | 160 | 210 |
|---|---|---|---|
| $i$ | 203 | 61.14 | 22.49 |
| $\ln i$ | 5.31 | 4.11 | 3.11 |

| $t$ | 275 | 320 | 390 |
|---|---|---|---|
| $i$ | 6.13 | 2.49 | 0.615 |
| $\ln i$ | 1.81 | 0.91 | −0.49 |

A graph of $\ln i$ against $t$ is shown in *Fig 7* and since a straight line results the law $i = Ie^{t/T}$ is verified.

**Fig 7**

Gradient of straight line, $\dfrac{1}{T} = \dfrac{AB}{BC} = \dfrac{5.30-1.30}{100-300} = \dfrac{4.0}{-200} = -0.02$

Hence $T = \dfrac{1}{-0.02} = \mathbf{-50}$

Selecting any point on the graph, say point D, where $t = 200$ and $\ln i = 3.31$, and substituting into $\ln i = \left(\dfrac{1}{T}\right)t + \ln I$

gives $\qquad 3.31 = \left(-\dfrac{1}{50}\right)(200) + \ln I$

from which $\ln I = 3.31 + 4.0 = 7.31$

and $\qquad I = \text{antilog } 7.31 \ (= e^{7.31}) = 1495$ or **1500**,

correct to 3 significant figures.

**Hence the law of the graph is** $i = 1500e^{-\frac{t}{50}}$.

---

## C. FURTHER PROBLEMS ON REDUCING NON-LINEAR LAWS TO LINEAR FORM

In *Problems 1 to 4*, $x$ and $y$ are two related variables and all other letters denote constants. For the stated laws to be verified it is necessary to plot graphs of the variables in a modified form. State for each (a) what should be plotted on the vertical axis, (b) what should be plotted on the horizontal axis, (c) the gradient and (d) the vertical axis intercept.

1. (i) $y = d + cx^2$ $\qquad\qquad$ [(i) (a) $y$; (b) $x^2$; (c) $c$; (d) $d$
   (ii) $y - a = b\sqrt{x}$ $\qquad\qquad$ (ii) (a) $y$; (b) $\sqrt{x}$; (c) $b$; (d) $a$]

2. (i) $y - e = \dfrac{f}{x}$ $\qquad\qquad$ [(i) (a) $y$; (b) $\dfrac{1}{x}$; (c) $f$; (d) $e$
   (ii) $y - cx = bx^2$ $\qquad\qquad$ (ii) (a) $\dfrac{y}{x}$; (b) $x$; (c) $b$; (d) $c$]

3. (i) $y = \dfrac{a}{x} + bx$ $\qquad\qquad$ [(i) (a) $\dfrac{y}{x}$; (b) $\dfrac{1}{x^2}$; (c) $a$; (d) $b$
   (ii) $y = ba^x$ $\qquad\qquad$ (ii) (a) $\lg y$; (b) $x$; (c) $\lg a$; (d) $\lg b$]

4. (i) $y = kx^l$ $\qquad\qquad$ [(i) (a) $\lg y$; (b) $\lg x$; (c) $l$; (d) $\lg k$
   (ii) $\dfrac{y}{m} = e^{nx}$ $\qquad\qquad$ (ii) (a) $\ln y$; (b) $x$; (c) $n$; (d) $\ln m$]

5. In an experiment the resistance of wire is measured for wires of different diameters with the following results.

   | R ohms | 1.64 | 1.14 | 0.89 | 0.76 | 0.63 |
   |---|---|---|---|---|---|
   | d mm | 1.10 | 1.42 | 1.75 | 2.04 | 2.56 |

   It is thought that $R$ is related to $d$ by the law $R = (a/d^2) + b$, where $a$ and $b$ are constants. Verify this and find the approximate values for $a$ and $b$. Determine the cross-sectional area needed for a resistance reading of 0.50 ohms.

   $[a = 1.5, b = 0.4; 11.78 \text{ mm}^2]$

6. Corresponding experimental values of two quantities $x$ and $y$ are given below.

   | x | 1.5 | 3.0 | 4.5 | 6.0 | 7.5 | 9.0 |
   |---|---|---|---|---|---|---|
   | y | 11.5 | 25.0 | 47.5 | 79.0 | 119.5 | 169.0 |

   By plotting a suitable graph verify that $y$ and $x$ are connected by a law of the form $y = kx^2 + c$, where $k$ and $c$ are constants. Determine the law of the graph and hence find the value of $x$ when $y$ is 60.0.

   $[y = 2x^2 + 7; 5.15]$

7  Experimental results of the safe load, $L$ kN, applied to girders of varying spans, $d$ m, are shown below.

| Span, $d$ m | 2.0 | 2.8 | 3.6 | 4.2 | 4.8 |
|---|---|---|---|---|---|
| Load, $L$ kN | 475 | 339 | 264 | 226 | 198 |

It is believed that the relationship between load and span is $L = c/d$, where $c$ is a constant. Determine (a) the value of constant $c$ and (b) the safe load for a span of 3.0 m.

[(a) 950; (b) 317 kN]

8  The following results give corresponding values of two quantities $x$ and $y$ which are believed to be related by a law of the form $y = ax^2 + bx$ where $a$ and $b$ are constants.

| $y$ | 33.86 | 55.54 | 72.80 | 84.10 | 111.4 | 168.1 |
|---|---|---|---|---|---|---|
| $x$ | 3.4 | 5.2 | 6.5 | 7.3 | 9.1 | 12.4 |

Verify the law and determine approximate values of $a$ and $b$. Hence determine (i) the value of $y$ when $x$ is 8.0 and (ii) the value of $x$ when $y$ is 146.5.

[$a = 0.4; b = 8.6$ (i) 94.4 (ii) 11.2]

9  The luminosity $I$ of a lamp varies with the applied voltage $V$ and the relationship between $I$ and $V$ is thought to be $I = kV^n$. Experimental results obtained are:

| $I$ candelas | 1.92 | 4.32 | 9.72 | 15.87 | 23.52 | 30.72 |
|---|---|---|---|---|---|---|
| $V$ volts | 40 | 60 | 90 | 115 | 140 | 160 |

Verify that the law is true and determine the law of the graph. Determine also the luminosity when 75 V is applied across the lamp.

[$I = 0.0012\ V^2$; 6.75 candelas]

10  The head of pressure $h$ and the flow velocity $v$ are measured and are believed to be connected by the law $v = ah^b$, where $a$ and $b$ are constants. The results are as shown below.

| $h$ | 10.6 | 13.4 | 17.2 | 24.6 | 29.3 |
|---|---|---|---|---|---|
| $v$ | 9.77 | 11.00 | 12.44 | 14.88 | 16.24 |

Verify that the law is true and determine values of $a$ and $b$.

[$a = 3.0; b = 0.5$]

11  Experimental values of $x$ and $y$ are measured as follows.

| $x$ | 0.4 | 0.9 | 1.2 | 2.3 | 3.8 |
|---|---|---|---|---|---|
| $y$ | 8.35 | 13.47 | 17.94 | 51.32 | 215.20 |

The law relating $x$ and $y$ is believed to be of the form $y = ab^x$, where $a$ and $b$ are constants. Determine the approximate values of $a$ and $b$. Hence find the value of $y$ when $x$ is 2.0 and the value of $x$ when $y$ is 100.

[$a = 5.7, b = 2.6$; 38.53; 3.0]

12 The activity of a mixture of radioactive isotope is believed to vary according to the law $R = R_0 t^{-c}$, where $R_0$ and $c$ are constants. Experimental results are shown below.

| R | 9.72 | 2.65 | 1.15 | 0.47 | 0.32 | 0.23 |
|---|---|---|---|---|---|---|
| t | 2 | 5 | 9 | 17 | 22 | 28 |

Verify that the law is true and determine approximate values of $R_0$ and $c$.
$$[R_0 = 26.0, c = 1.42]$$

13 Determine the law of the form $y = ae^{kx}$ which relates the following values.

| y | 0.0306 | 0.285 | 0.841 | 5.21 | 173.2 | 1181 |
|---|---|---|---|---|---|---|
| x | −4.0 | 5.3 | 9.8 | 17.4 | 32.0 | 40.0 |

$$[y = 0.08\, e^{0.24x}]$$

14 The tension $T$ in a belt passing round a pulley wheel and in contact with the pulley over an angle of $\theta$ radians is given by $T = T_0 e^{\mu\theta}$, where $T_0$ and $\mu$ are constants. Experimental results obtained are:

| T newtons | 47.9 | 52.8 | 60.3 | 70.1 | 80.9 |
|---|---|---|---|---|---|
| θ radians | 1.12 | 1.48 | 1.97 | 2.53 | 3.06 |

Determine approximate values of $T_0$ and $\mu$. Hence find the tension when $\theta$ is 2.25 radians and the value of $\theta$ when the tension is 50.0 newtons.
$$[T_0 = 35.4 \text{ N}, \mu = 0.27; 65.0 \text{ N}; 1.28 \text{ radians}]$$

# 15 Graphs with logarithmic scales

## A  MAIN POINTS CONCERNED WITH GRAPHS HAVING LOGARITHMIC SCALES

1   (i) Graph paper is available where the scale markings along the horizontal and vertical axes are proportional to the logarithms of the numbers. Such graph paper is called **log-log graph paper**.

    Fig 1

   (ii) A logarithmic scale is shown in *Fig 1* where the distances between, say 1 and 2 is proportional to lg 2−lg 1, i.e. 0.3010 of the total distance from 1 to 10. Similarly, the distance between 7 and 8 is proportional to lg 8−lg 7, i.e. 0.057 99 of the total distance from 1 to 10. Thus the distance between markings progressively decreases as the numbers increase from 1 to 10.

   (iii) With log-log graph paper the scale markings are from 1 to 9, and this pattern can be repeated several times. The number of times the pattern of markings is repeated on an axis signifies the number of **cycles**. When the vertical axis has, say, 3 sets of values from 1 to 9 and the horizontal axis has 2 sets of values from 1 to 9, then this log-log graph paper is called 'log 3 cycle × 2 cycle' (see *Fig 2*). Many different arrangements are available ranging from 'log 1 cycle × 1 cycle' through to 'log 5 cycle × 5 cycle'.

   (iv) To depict a set of values, say, from 0.4 to 161 on an axis of log-log graph paper, 4 cycles are required, from 0.1 to 1, 1 to 10, 10 to 100 and 100 to 1000.

2   (i) **To express graphs of the form $y = ax^n$ in linear form**
    Taking logarithms to a base of 10 of both sides of $y = ax^n$ gives:
    $$\lg y = \lg (ax^n) = \lg a + \lg x^n$$
    i.e. $\lg y = n \lg x + \lg a$
    which compares with $Y = mX + c$.
    Thus, by plotting lg $y$ vertically against lg $x$ horizontally, a straight line results, i.e. the equation $y = ax^n$ is reduced to linear form. With log-log graph paper available $x$ and $y$ may be plotted directly, without having to firstly determine their logarithms, as shown in chapter 14. (See *Problems 1 to 3*.)

   (ii) **To express graphs of the form $y = ab^x$ in linear form**
    Taking logarithms to a base of 10 of both sides of $y = ab^x$ gives:
    $$\lg y = \lg (ab^x) = \lg a + \lg b^x = \lg a + x \lg b$$
    i.e. $\lg y = (\lg b)x + \lg a$
    which compares with $Y = mX + c$
    Thus, by plotting lg $y$ vertically against $x$ horizontally a straight line results, i.e. the graph $y = ab^x$ is reduced to linear form. In this case, graph paper having a linear horizontal scale and a logarithmic vertical scale may be used. This type of graph paper is called **log-linear graph paper**, and is specified by the number

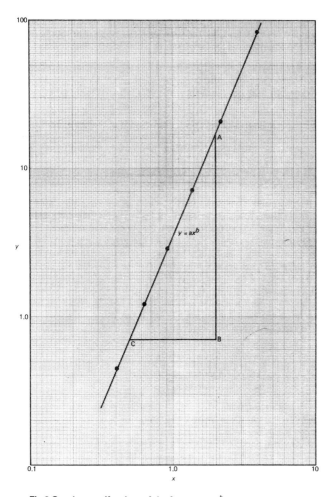

**Fig 2 Graph to verify a law of the form** $y = ax^b$

of cycles on the logarithmic scale. For example, graph paper having 3 cycles on the logarithmic scale is called 'log 3 cycle × linear' graph paper. (See *Problem 4*.)

(iii) **To express graphs of the form** $y = ae^{kx}$ **in linear form**

Taking logarithms to a base of $e$ of both sides of $y = ae^{kx}$ gives:
$$\ln y = \ln (ae^{kx}) = \ln a + \ln e^{kx} = \ln a + kx \ln e$$
i.e. $\ln y = kx + \ln a$, (since $\ln e = 1$),

which compares with $Y = mX + c$.

Thus, by plotting $\ln y$ vertically against $x$ horizontally, a straight line results, i.e. the equation $y = ae^{kx}$ is reduced to linear form. Since $\ln y = 2.3026 \lg y$, i.e. $\ln y = $ (a constant)($\lg y$), the same log-linear graph paper can be used for Naperian logarithms as for logarithms to a base of 10. (See *Problems 5 and 6*.)

## B. WORKED PROBLEMS ON GRAPHS HAVING LOGARITHMIC SCALES

*Problem 1* Experimental values of two related quantities $x$ and $y$ are shown below:

| $x$ | 0.41 | 0.63 | 0.92 | 1.36 | 2.17 | 3.95 |
|---|---|---|---|---|---|---|
| $y$ | 0.45 | 1.21 | 2.89 | 7.10 | 20.79 | 82.46 |

The law relating $x$ and $y$ is believed to be $y = ax^b$, where $a$ and $b$ are constants. Verify that this law is true and determine the approximate values of $a$ and $b$.
If $y = ax^b$ then $\lg y = b \lg x + \lg a$, from para. 2(i), which is of the form $Y = mX + c$, showing that to produce a straight line graph $\lg y$ is plotted vertically against $\lg x$ horizontally. $x$ and $y$ may be plotted directly on to log-log graph paper as shown in *Fig 2*. The values of $y$ range from 0.45 to 82.46 and 3 cycles are needed, (i.e. 0.1 to 1, 1 to 10 and 10 to 100). The values of $x$ range from 0.41 to 3.95 and 2 cycles are needed (i.e. 0.1 to 1 and 1 to 10). Hence 'log 3 cycle × 2 cycle' is used as shown in *Fig 2* where the axes are marked and the points plotted. Since the points lie on a straight line the law $y = ax^b$ is verified.

**To evaluate constants $a$ and $b$:**

*Method 1.* Any two points on the straight line, say points A and C are selected, and AC and BC are measured (say in centimetres).

Then, gradient, $b = \dfrac{AC}{BC} = \dfrac{11.5 \text{ units}}{5 \text{ units}} = 2.3$

Since $\lg y = b \lg x + \lg a$, when $x = 1$, $\lg x = 0$ and $\lg y = \lg a$.
The straight line crosses the ordinate $x = 1.0$ at $y = 3.5$
Hence $\lg a = \lg 3.5$, i.e. $a = \mathbf{3.5}$

*Method 2.* Any two points on the straight line, say points A and C, are selected.
A has co-ordinates (2, 17.25) and C has co-ordinates (0.5, 0.7).
Since $y = ax^b$ then $17.25 = a(2)^b$ (1)
and $0.7 = a(0.5)^b$ (2)

i.e., two simultaneous equations are produced and may be solved for $a$ and $b$.
Dividing equation (1) by equation (2) to eliminate $a$ gives $\dfrac{17.25}{0.7} = \dfrac{(2)^b}{(0.5)^b} = \left(\dfrac{2}{0.5}\right)^b$
i.e. $24.643 = (4)^b$.
Taking logarithms of both sides gives $\lg 24.643 = b \lg 4$
i.e. $b = \dfrac{\lg 24.643}{\lg 4} = 2.3$, correct to 2 significant figures.

Substituting $b = 2.3$ in equation (1) gives: $17.25 = a(2)^{2.3}$

i.e. $a = \dfrac{17.25}{(2)^{2.3}} = \dfrac{17.25}{4.925} = \mathbf{3.5}$, correct to 2 significant figures.

**Hence the law of the graph is $y = 3.5x^{2.3}$**

*Problem 2* The power dissipated by a resistor was measured for varying values of current flowing in the resistor and the results are as shown:

| Current, $I$ amperes | 1.4 | 4.7 | 6.8 | 9.1 | 11.2 | 13.1 |
|---|---|---|---|---|---|---|
| Power, $P$ watts | 49 | 552 | 1156 | 2070 | 3136 | 4290 |

Prove that the law relating current and power is of the form $P = RI^n$, where $R$ and $n$ are constants, and determine the law. Hence calculate the power when the current is 12 amperes and the current when the power is 1000 W.

Since $P = RI^n$ then $\lg P = n \lg I + \lg R$, (from para. 2(i)), which is of the form $Y = mX + c$, showing that to produce a straight line graph $\lg P$ is plotted vertically against $\lg I$ horizontally. Power values range from 49 to 4290, hence 3 cycles of log-log graph paper are needed (10 to 100, 100 to 1000 and 1000 to 10 000).

Current values range from 1.4 to 11.2, hence 2 cycles of log-log graph paper are needed (1 to 10 and 10 to 100).

Thus 'log 3 cycles × 2 cycles' is used as shown in *Fig 3* (or, if not available, graph paper having a larger number of cycles per axis can be used). The co-ordinates are plotted and a straight line results which proves that the law relating current and power is of the form $P = RI^n$.

Gradient of straight line $n = \dfrac{AB}{BC} = \dfrac{14 \text{ units}}{7 \text{ units}} = 2$

At point C, $I = 2$ and $P = 100$. Substituting these values into $P = RI^n$ gives: $100 = R(2)^2$. Hence $R = 100/(2)^2 = 25$ which may have been found from the intercept on the $I = 1.0$ axis in *Fig 3*. **Hence the law of the graph is $P = 25I^2$**.
When current $I = 12$, power $P = 25(12)^2 = $ **3600 watts** (which may be read from the graph).
When power $P = 1000$, $1000 = 25I^2$

Hence $\quad I^2 = \dfrac{1000}{25} = 40$

from which, $I = \sqrt{40} = $ **6.32 A**

---

*Problem 3* The pressure $p$ and volume $v$ of a gas are believed to be related by a law of the form $p = cv^n$, where $c$ and $n$ are constants. Experimental values of $p$ and corresponding values of $v$ obtained in a laboratory are:

| $p$ pascals | $2.28 \times 10^5$ | $8.04 \times 10^5$ | $2.03 \times 10^6$ | $5.05 \times 10^6$ | $1.82 \times 10^7$ |
|---|---|---|---|---|---|
| $v$ m$^3$ | $3.2 \times 10^{-2}$ | $1.3 \times 10^{-2}$ | $6.7 \times 10^{-3}$ | $3.5 \times 10^{-3}$ | $1.4 \times 10^{-3}$ |

Verify that the law is true and determine approximate values of $c$ and $n$.

Since $p = cv^n$, then $\lg p = n \lg v + \lg c$, which is of the form $Y = mX + c$, showing that to produce a straight line graph $\lg p$ is plotted vertically against $\lg v$ horizontally. The co-ordinates are plotted on 'log 3 cycle × 2 cycle' graph paper as shown

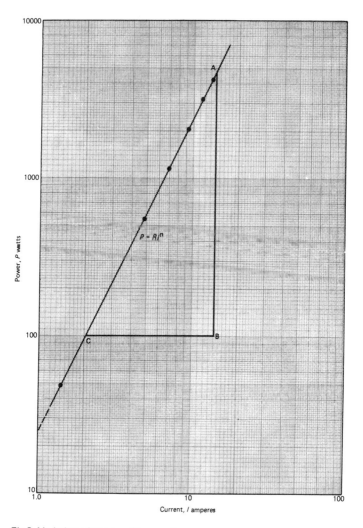

Fig 3 Variation of power with current

in *Fig 4*. With the data expressed in standard form, the axes are marked in standard form also. Since a straight line results the law $p = cv^n$ is verified.

The straight line has a negative gradient and the value of the gradient is given by

$\dfrac{AB}{BC} = \dfrac{14 \text{ units}}{10 \text{ units}} = 1.4$. Hence $n = -1.4$

Selecting any point on the straight line, say point C, having co-ordinates

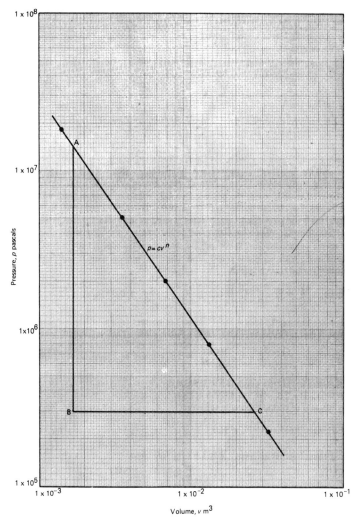

**Fig 4 Variation of pressure with volume**

$(2.63 \times 10^{-2}, 3 \times 10^5)$, and substituting these values in $p = cv^n$ gives:

$$3 \times 10^5 = c(2.63 \times 10^{-2})^{-1.4}$$

Hence $\quad c = \dfrac{3 \times 10^5}{(2.63 \times 10^{-2})^{-1.4}} = \dfrac{3 \times 10^5}{(0.0263)^{-1.4}} = \dfrac{3 \times 10^5}{1.63 \times 10^2}$

$= 1840$, correct to 3 significant figures.

**Hence the law of the graph is $p = 1840v^{-1.4}$ or $pv^{1.4} = 1840$**

*Problem 4* Experimental values of quantities $x$ and $y$ are believed to be related by a law of the form $y = ab^x$, where $a$ and $b$ are constants. The values of $x$ and corresponding values of $y$ are:

| $x$ | 0.7 | 1.4 | 2.1 | 2.9 | 3.7 | 4.3 |
|---|---|---|---|---|---|---|
| $y$ | 18.4 | 45.1 | 111 | 308 | 858 | 1850 |

Verify the law and determine the approximate values of $a$ and $b$. Hence evaluate (i) the value of $y$ when $x$ is 2.5 and (ii) the value of $x$ when $y$ is 1200.

Since $y = ab^x$ then $\lg y = (\lg b)x + \lg a$, (from para. 2(ii)), which is of the form $Y = mX + c$, showing that to produce a straight line graph $\lg y$ is plotted vertically against $x$ horizontally. Using log-linear graph paper, values of $x$ are marked on the horizontal scale to cover the range 0.7 to 4.3. Values of $y$ range from 18.4 to 1850 and 3 cycles are needed (i.e. 10 to 100, 100 to 1000 and 1000 to 10 000). Thus 'log 3 cycles × linear' graph paper is used as shown in *Fig 5*. A straight line is drawn through the co-ordinates, hence the law $y = ab^x$ is verified. Gradient of straight line, $\lg b = AB/BC$. Direct measurement (say in centimetres) is not made with log-linear graph paper since the vertical scale is logarithmic and the horizontal scale is linear.

Hence $\dfrac{AB}{BC} = \dfrac{\lg 1000 - \lg 100}{3.82 - 2.02} = \dfrac{3-2}{1.80} = \dfrac{1}{1.80} = 0.5556$

Here $b = $ antilog $0.5556 (= 10^{0.5556}) = \mathbf{3.6}$, correct to 2 significant figures.
Point A has co-ordinates (3.82, 1000). Substituting these values into $y = ab^x$ gives:

$1000 = a(3.6)^{3.82}$, i.e. $a = \dfrac{1000}{(3.6)^{3.82}} = \mathbf{7.5}$, correct to 2 significant figures.

**Hence the law of the graph is $y = 7.5(3.6)^x$**
(i) When $x = 2.5$, $y = 7.5(3.6)^{2.5} = \mathbf{184}$
(ii) When $y = 1200$, $1200 = 7.5(3.6)^x$

Hence $(3.6)^x = \dfrac{1200}{7.5} = 160$

Taking logarithms gives $x \lg 3.6 = \lg 160$

i.e. $x = \dfrac{\lg 160}{\lg 3.6} = \dfrac{2.2041}{0.5563} = \mathbf{3.96}$

*Problem 5* The data given below is believed to be related by a law of the form $y = ae^{kx}$, where $a$ and $b$ are constants. Verify that the law is true and determine approximate values of $a$ and $b$. Also determine the value of $y$ when $x$ is 3.8 and the value of $x$ when $y$ is 85.

| $x$ | −1.2 | 0.38 | 1.2 | 2.5 | 3.4 | 4.2 | 5.3 |
|---|---|---|---|---|---|---|---|
| $y$ | 9.3 | 22.2 | 34.8 | 71.2 | 117 | 181 | 332 |

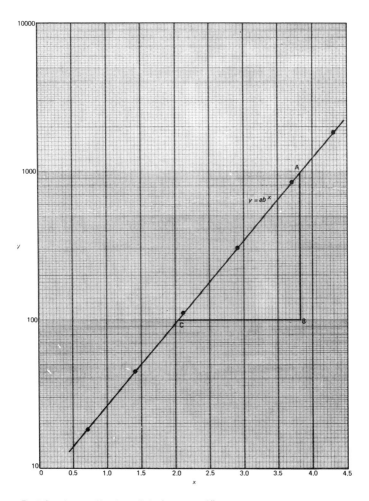

**Fig 5 Graph to verify a law of the form $y = ab^x$**

Since $y = ae^{kx}$ then $\ln y = kx + \ln a$, (from para. 2(iii)), which is of the form $Y = mX + c$, showing that to produce a straight line graph $\ln y$ is plotted vertically against $x$ horizontally. The value of $y$ ranges from 9.3 to 332 hence 'log 3 cycle × lines' graph paper is used. The plotted co-ordinates are shown in *Fig 6* and since a straight line passes through the points the law $y = ae^{kx}$ is verified.

Gradient of straight line, $k = \dfrac{AB}{BC} = \dfrac{\ln 100 - \ln 10}{3.12 - (-1.08)} = \dfrac{2.3026}{4.20}$

$\qquad\qquad\qquad\quad = 0.55$, correct to 2 significant figures.

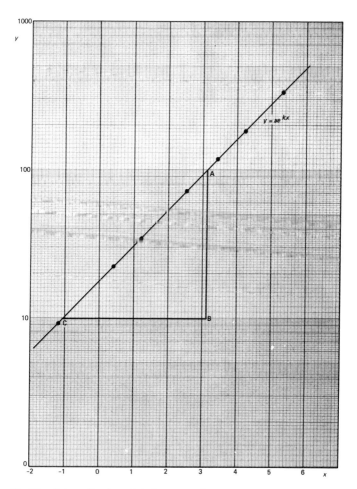

**Fig 6** Graph to verify a law of the form $y = ae^{kx}$

Since $\ln y = kx + \ln a$, when $x = 0$, $\ln y = \ln a$, i.e. $y = a$
The vertical axis intercept value at $x = 0$ is 18, hence $a = \mathbf{18}$
The law of the graph is thus $y = \mathbf{18}e^{\mathbf{0.55}x}$
When $x$ is 3.8, $y = 18e^{0.55(3.8)} = 18e^{2.09} = 18(8.0849) = \mathbf{146}$
When $y$ is 85, $85 = 18e^{0.55x}$. Hence $e^{0.55x} = \dfrac{85}{18} = 4.7222$
and $\quad 0.55x = \ln 4.7222 = 1.5523$. Hence $x = \dfrac{1.5523}{0.55} = \mathbf{2.82}$

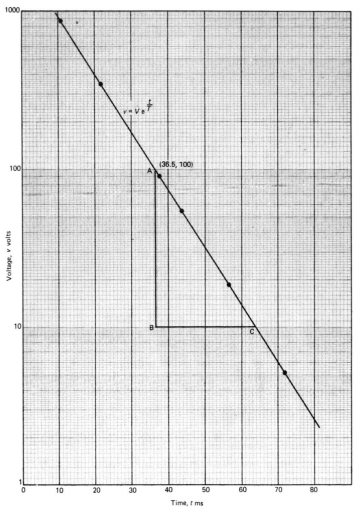

**Fig 7** Variation of voltage with time

*Problem 6* The voltage, $v$ volts, across an inductor is believed to be related to time, $t$ ms, by the law $v = Ve^{t/T}$, where $V$ and $T$ are constants. Experimental results obtained are:

| $v$ volts | 883 | 347 | 90 | 55.5 | 18.6 | 5.2 |
|---|---|---|---|---|---|---|
| $t$ ms | 10.4 | 21.6 | 37.8 | 43.6 | 56.7 | 72.0 |

Show that the law relating voltage and time is as stated and determine the approximate values of $V$ and $T$. Find also the value of voltage after 25 ms and the time when the voltage is 30.0 V.

Since $v = Ve^{t/T}$ then $\ln v = \dfrac{1}{T}t + \ln V$, from para 2(iii), which is of the form $Y = mX + c$.

Using 'log 3 cycle × linear' graph paper, the points are plotted as shown in *Fig 7*. Since the points are joined by a straight line the law $v = Ve^{t/T}$ is verified.

Gradient of straight line, $\dfrac{1}{T} = \dfrac{AB}{BC} = \dfrac{\ln 100 - \ln 10}{36.5 - 64.2} = \dfrac{2.3026}{-27.7}$

Hence $T = \dfrac{-27.7}{2.3026} = -12.0$, correct to 3 significant figures.

Since the straight line does not cross the vertical axis at $t = 0$ in *Fig 7*, the value of $V$ is determined by selecting any point, say A, having co-ordinates (36.5, 100) and substituting these values into $v = Ve^{t/T}$.

Thus $100 = Ve^{-\frac{36.5}{12.0}}$

i.e. $V = \dfrac{100}{e^{-\frac{36.5}{12.0}}} = 2090$ volts, correct to 3 significant figures.

Hence the law of the graph is $v = 2090e^{-\frac{t}{12.0}}$

When time $t = 25$ ms, voltage $v = 2090e^{-\frac{25}{12.0}} = 260$ V

When the voltage is 30.0 volts, $30.0 = 2090e^{-\frac{t}{12.0}}$

Hence $e^{-\frac{t}{12.0}} = \dfrac{30}{2090}$ and $e^{\frac{t}{12.0}} = \dfrac{2090}{30} = 69.67$

Taking Naperian logarithms gives: $\dfrac{t}{12.0} = \ln 69.67 = 4.2438$, from which, time $t = (12.0)(4.2438) = 50.9$ ms

## C. FURTHER PROBLEMS ON GRAPHS HAVING LOGARITHMIC SCALES

1. Quantities $x$ and $y$ are believed to be related by a law of the form $y = ax^n$, where $a$ and $n$ are constants. Experimental values of $x$ and corresponding values of $y$ are:

| $x$ | 0.8 | 2.3 | 5.4 | 11.5 | 21.6 | 42.9 |
|---|---|---|---|---|---|---|
| $y$ | 8 | 54 | 250 | 974 | 3028 | 10 410 |

Show that the law is true and determine the values of $a$ and $n$. Hence determine the value of $y$ when $x$ is 7.5 and the value of $x$ when $y$ is 5000.

$[a = 12, n = 1.8; 451; 28.5]$

2. Show from the following results of voltage $V$ and admittance $Y$ of an electrical circuit that the law connecting the quantities is of the form $V = kY^n$, and determine the values of $k$ and $n$.

| Voltage, $V$ volts | 2.88 | 2.05 | 1.60 | 1.22 | 0.96 |
|---|---|---|---|---|---|
| Admittance, $Y$ siemens | 0.52 | 0.73 | 0.94 | 1.23 | 1.57 |

$$[k = 1.5; n = -1]$$

3 Quantities $x$ and $y$ are believed to be related by a law of the form $y = mn^x$. The values of $x$ and corresponding values of $y$ are:

| $x$ | 0 | 0.5 | 1.0 | 1.5 | 2.0 | 2.5 | 3.0 |
|---|---|---|---|---|---|---|---|
| $y$ | 1.0 | 3.2 | 10 | 31.6 | 100 | 316 | 1000 |

Verify the law and find the values of $m$ and $n$.

$$[m = 1; n = 10]$$

4 Experimental values of $p$ and corresponding values of $q$ are shown below.

| $p$ | −13.2 | −27.9 | −62.2 | −383.2 | −1581 | −2931 |
|---|---|---|---|---|---|---|
| $q$ | 0.30 | 0.75 | 1.23 | 2.32 | 3.17 | 3.54 |

Show that the law relating $p$ and $q$ is $p = ab^q$, where $a$ and $b$ are constants. Determine (i) value of $a$ and $b$, and state the law, (ii) the value of $p$ when $q$ is 2.0 and (iii) the value of $q$ when $p$ is −2000.

$$\begin{bmatrix} \text{(i) } a = -8, b = 5.3, p = -8(5.3)^q; \\ \text{(ii) } -224.7; \text{(iii) } 3.31 \end{bmatrix}$$

5 Atmospheric pressure $p$ is measured at varying altitudes $h$ and the results are as shown below:

| Altitude, $h$ m | 500 | 1500 | 3000 | 5000 | 8000 |
|---|---|---|---|---|---|
| pressure, $p$ cm | 73.39 | 68.42 | 61.60 | 53.56 | 43.41 |

Show that the quantities are related by the law $p = ae^{kh}$, where $a$ and $k$ are constants. Determine the values of $a$ and $k$ and state the law. Find also the atmospheric pressure at 10 000 m.

$$[a = 76, k = -7 \times 10^{-5}, p = 76\, e^{-7 \times 10^{5} h}, 37.74 \text{ cm}]$$

6 At particular times, $t$ minutes, measurements are made of the temperature, $\theta°C$ of a cooling liquid and the following results are obtained:

| Temperature $\theta°C$ | 92.2 | 55.9 | 33.9 | 20.6 | 12.5 |
|---|---|---|---|---|---|
| Time $t$ minutes | 10 | 20 | 30 | 40 | 50 |

Prove that the quantities follow a law of the form $\theta = \theta_0 e^{kt}$, where $\theta_0$ and $k$ are constants, and determine the approximate values of $\theta_0$ and $k$.

$$[\theta_0 = 152; k = -0.05]$$

7−12 See *Problems 9 to 14* on page 142 which may all be solved using logarithmic graph paper.

# 16 Graphical solution of equations

## A. MAIN POINTS CONCERNED WITH THE GRAPHICAL SOLUTION OF EQUATIONS

1   **Linear simultaneous equations** in two unknowns may be solved graphically by:
    (i) plotting the two straight lines on the same axes, and
    (ii) noting their point of intersection.
    The co-ordinates of the point of intersection give the required solution. (See *Problems 1 and 2*.)
2   A general **quadratic equation** is of the form $y = ax^2 + bx + c$, where $a$, $b$ and $c$ are constants and $a$ is not equal to zero. A graph of a quadratic equation always produces a shape called a **parabola**.
3   The gradient of the curve between 0 and A and between B and C in *Fig 1* is positive, whilst the gradient between A and B is negative. Points such as A and B are called **turning points**. At A the gradient is zero and, as $x$ increases, the gradient of the curve changes from positive just before A to negative just after. Such a point is called a **maximum value**. At B the gradient is also zero and, as $x$ increases, the gradient of the curve changes from negative just before B to positive just after. Such a point is called a **minimum value**.

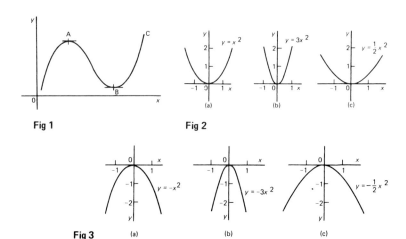

**Fig 1**   **Fig 2**

**Fig 3**   (a)   (b)   (c)

4  **Quadratic graphs.** (i) $y = ax^2$
   Graphs of $y = x^2$, $y = 3x^2$ and $y = \frac{1}{2}x^2$ are shown in
   *Fig 2*. All have minimum values at the origin (0,0).
   Graphs of $y = -x^2$, $y = -3x^2$ and $y = -\frac{1}{2}x^2$ are shown in *Fig 3*. All have
   maximum values at the origin (0,0).
   When $y = ax^2$, (a) curves are symmetrical about the $y$-axis,
   (b) the magnitude of '$a$' affects the gradient of the curve,
   and (c) the sign of '$a$' determines whether it has a maximum or minimum value.

(ii) $y = ax^2 + c$
   Graphs of $y = x^2 + 3$, $y = x^2 - 2$, $y = -x^2 + 2$ and $y = -2x^2 - 1$ are shown in *Fig 4*.

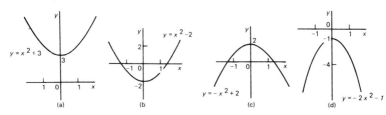

**Fig 4**

When $y = ax^2 + c$: (a) curves are symmetrical about the $y$-axis,
(b) the magnitude of '$a$' affects the gradient of the curve,
and (c) the constant '$c$' is the $y$-axis intercept.

(iii) $y = ax^2 + bx + c$.
   Whenever '$b$' has a value other than zero the curve is displaced to the right or left of the $y$-axis. When $b/a$ is positive, the curve is displaced $b/2a$ to the left of the $y$-axis, as shown in *Fig 5(a)*. When $b/a$ is negative the curve is displaced $b/2a$ to the right of the $y$-axis, as shown in *Fig 5(b)*.

5  **Quadratic equations** of the form $ax^2 + bx + c = 0$ may be solved graphically by:
   (i) plotting the graph $y = ax^2 + bx + c$, and
   (ii) noting the points of intersection on the $x$-axis (i.e. where $y = 0$).

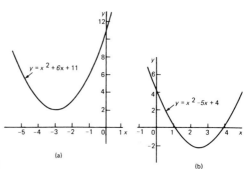

**Fig 5**

The $x$ values of the points of intersection give the required solutions since at these points both $y = 0$ and $ax^2+bx+c = 0$. The number of solutions, or roots of a quadratic equation depends on how many times the curve cuts the $x$-axis and there can be no real roots (as in *Fig 5(a)*) or one root (as in *Figs 2 and 3*) or two roots (as in *Fig 5(b)*). (See *Problems 3 to 6.*)

6  The solution of **linear and quadratic equations simultaneously** may be achieved graphically by: (i) plotting the straight line and parabola on the same axes, and (ii) noting the points of intersection. The co-ordinates of the points of intersection give the required solutions. (See *Problem 7.*)

7  A **cubic equation** of the form $ax^3+bx^2+cx+d = 0$ may be solved graphically by: (i) plotting the graph $y = ax^3+bx^2+cx+d$, and (ii) noting the points of intersection on the $x$-axis, (i.e. where $y = 0$). The $x$-values of the points of intersection give the required solution since at these points both $y = 0$ and $ax^3+bx^2+cx+d = 0$.

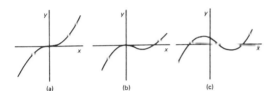

**Fig 6**

The number of solutions, or roots of a cubic equation depends on how many times the curve cuts the $x$-axis and there can be one, two or three possible roots, as shown in *Fig 6*. (See *Problems 8 and 9.*)

## B. WORKED PROBLEMS ON THE GRAPHICAL SOLUTION OF EQUATIONS

*Problem 1* Solve graphically the simultaneous equations: $2x-y = 4$
$\qquad\qquad\qquad\qquad\qquad\qquad\qquad\qquad\qquad\qquad\quad x+y = 5$

Rearranging each equation into $y = mx+c$ form gives:

$y = 2x-4$  (1)
$y = -x+5$  (2)

Only three co-ordinates need be calculated for each graph since both are straight lines.

| $x$ | 0 | 1 | 2 |
|---|---|---|---|
| $y = 2x-4$ | $-4$ | $-2$ | 0 |

| $x$ | 0 | 1 | 2 |
|---|---|---|---|
| $y = -x+5$ | 5 | 4 | 3 |

Each of the graphs are plotted as shown in *Fig 7*. The point of intersection is at (3,2) and since this is the only point which lies simultaneously on both lines then $x = 3, y = 2$ is the solution of the simultaneous equations.

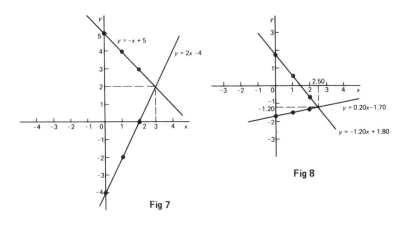

Fig 7

Fig 8

*Problem 2* Solve graphically the equations $1.20x + y = 1.80$
$x - 5.0y = 8.50$

Rearranging each equation into $y = mx + c$ form gives:

$$y = -1.20x + 1.80 \quad (1)$$

$$y = \frac{x}{5.0} - \frac{8.5}{5.0}$$

i.e. $y = 0.20x - 1.70 \quad (2)$

Three co-ordinates are calculated for each equation as shown below.

| $x$ | 0 | 1 | 2 | $x$ | 0 | 1 | 2 |
|---|---|---|---|---|---|---|---|
| $y = -1.20x + 1.80$ | 1.80 | 0.60 | −0.60 | $y = 0.20x - 1.70$ | −1.70 | −1.50 | −1.30 |

The two lines are plotted as shown in *Fig 8*. The point of intersection is $(2.50, -1.20)$. Hence the solution of the simultaneous equations is $x = 2.50$, $y = -1.20$ (It is sometimes useful to initially sketch the two straight lines to determine the region where the point of intersection is. Then, for greater accuracy, a graph having a smaller range of values can be drawn to 'magnify' the point of intersection.)

*Problem 3* Solve the quadratic equation $4x^2 + 4x - 15 = 0$ graphically given that the solutions lie in the range $x = -3$ to $x = 2$. Determine also the co-ordinates and nature of the turning point of the curve.

Let $y = 4x^2 + 4x - 15$. A table of values is drawn up as shown below.

| $x$ | $-3$ | $-2$ | $-1$ | $0$ | $1$ | $2$ |
|---|---|---|---|---|---|---|
| $4x^2$ | 36 | 16 | 4 | 0 | 4 | 16 |
| $4x$ | $-12$ | $-8$ | $-4$ | 0 | 4 | 8 |
| $-15$ | $-15$ | $-15$ | $-15$ | $-15$ | $-15$ | $-15$ |
| $y = 4x^2+4x-15$ | 9 | $-7$ | $-15$ | $-15$ | $-7$ | 9 |

A graph of $y = 4x^2+4x-15$ is shown in *Fig 9*. The only points where $y = 4x^2+4x-15$ and $y = 0$ are the points marked A and B. This occurs at $x = -2.5$ and $x = 1.5$ and these are the solutions of the quadratic equation $4x^2+4x-15 = 0$. (By substituting $x = -2.5$ and $x = 1.5$ into the original equation

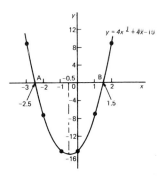

**Fig 9**

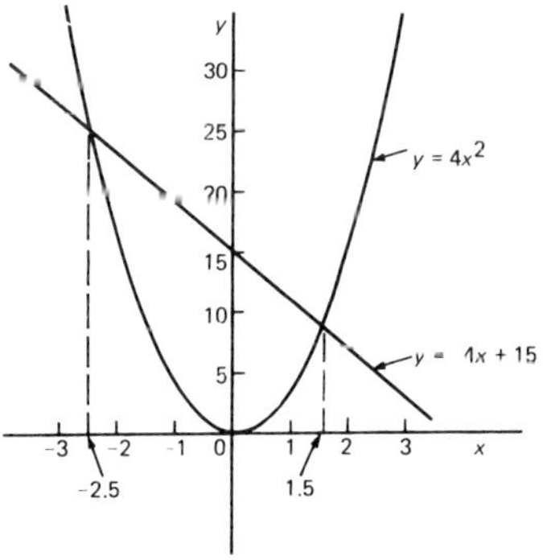

**Fig 10**

the solutions may be checked.) The curve has a turning point at $(-0.5, -16)$ and the nature of the point is a **minimum**.

An alternative graphical method of solving $4x^2+4x-15 = 0$ is to rearrange the equation as $4x^2 = -4x+15$ and then plot two separate graphs—in this case $y = 4x^2$ and $y = -4x+15$. Their points of intersection give the roots of equation $4x^2 = -4x+15$, i.e. $4x^2+4x-15 = 0$. This is shown in *Fig 10*, where the roots are $x = -2.5$ and $x = 1.5$ as before.

*Problem 4* Solve graphically the quadratic equation $-5x^2+9x+7.2 = 0$ given that the solutions lie between $x = -1$ and $x = 3$. Determine also the co-ordinates of the turning point and state its nature.

Let $y = -5x^2+9x+7.2$. A table of values is drawn up as shown opposite. A graph of $y = -5x^2+9x+7.2$ is shown plotted in *Fig 11*. The graph crosses the $x$-axis (i.e., where $y = 0$) at $x = -0.6$ and $x = 2.4$ and these are the solutions of the quadratic equation $-5x^2+9x+7.2 = 0$. The turning point is a **maximum** having co-ordinates **(0.9, 11.25)**.

| $x$ | −1 | −0.5 | 0 | 1 | 2 | 2.5 | 3 |
|---|---|---|---|---|---|---|---|
| $-5x^2$ | −5 | −1.25 | 0 | −5 | −20 | −31.25 | −45 |
| $+9x$ | −9 | −4.5 | 0 | 9 | 18 | 22.5 | 27 |
| $+7.2$ | 7.2 | 7.2 | 7.2 | 7.2 | 7.2 | 7.2 | 7.2 |
| $y = -5x^2 + 9x + 7.2$ | −6.8 | 1.45 | 7.2 | 11.2 | 5.2 | −1.55 | −10.8 |

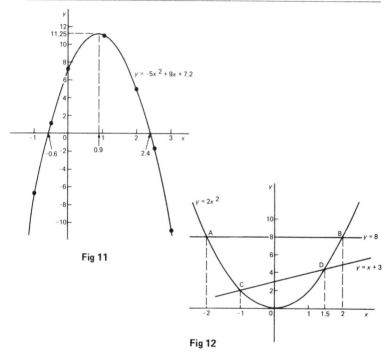

Fig 11

Fig 12

*Problem 5* Plot a graph of $y = 2x^2$ and hence solve the equations
(a) $2x^2 - 8 = 0$ and (b) $2x^2 - x - 3 = 0$.

A graph of $y = 2x^2$ is shown in *Fig 12*.
(a) Rearranging $2x^2 - 8 = 0$ gives $2x^2 = 8$ and the solution of this equation is obtained from the points of intersection of $y = 2x^2$ and $y = 8$, i.e. at co-ordinates (−2, 8) and (2, 8), shown as A and B respectively in *Fig 12*. Hence the solutions of $2x^2 - 8 = 0$ are $x = -2$ **and** $x = +2$.
(b) Rearranging $2x^2 - x - 3 = 0$ gives $2x^2 = x + 3$ and the solution of this equation is obtained from the points of intersection of $y = 2x^2$ and $y = x + 3$, i.e., at C and D in *Fig 12*. Hence the solutions of $2x^2 - x - 3 = 0$ are $x = -1$ **and** $x = 1.5$.

*Problem 6* Plot the graph of $y = -2x^2 + 3x + 6$ for values of $x$ from $x = -2$ to $x = 4$. Use the graph to find the roots of the following equations:
(a) $-2x^2 + 3x + 6 = 0$; (b) $-2x^2 + 3x + 2 = 0$; (c) $-2x^2 + 3x + 9 = 0$;
(d) $-2x^2 + x + 5 = 0$.

A table of values is drawn up as shown below.

| $x$ | $-2$ | $-1$ | $0$ | $1$ | $2$ | $3$ | $4$ |
|---|---|---|---|---|---|---|---|
| $-2x^2$ | $-8$ | $-2$ | $0$ | $-2$ | $-8$ | $-18$ | $-32$ |
| $+3x$ | $-6$ | $-3$ | $0$ | $3$ | $6$ | $9$ | $12$ |
| $+6$ | $6$ | $6$ | $6$ | $6$ | $6$ | $6$ | $6$ |
| $y$ | $-8$ | $1$ | $6$ | $7$ | $4$ | $-3$ | $-14$ |

A graph of $y = -2x^2 + 3x + 6$ is shown in Fig 13.
(a) The parabola $y = -2x^2 + 3x + 6$ and the straight line $y = 0$ intersect at A and D, where $x = -1.13$ and $x = 2.63$ and these are the roots of the equation $-2x^2 + 3x + 6 = 0$

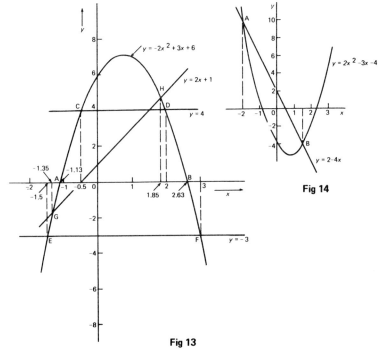

Fig 13

Fig 14

(b) Comparing $y = -2x^2+3x+6$ (1) with $0 = -2x^2+3x+2$ (2) shows that if 4 is added to both sides of equation (2), the right hand side of both equations will be the same. Hence $4 = -2x^2+3x+6$. The solution of this equation is found from the points of intersection of the line $y = 4$ and the parabola $y = -2x^2+3x+6$, i.e. points C and D in *Fig 13*. Hence the roots of $-2x^2+3x+2 = 0$ are $x = -0.5$ and $x = 2$

(c) $-2x^2+3x+9 = 0$ may be rearranged as $-2x^2+3x+6 = -3$, and the solution of this equation is obtained from the points of intersection of the line $y = -3$ and the parabola $y = -2x^2+3x+6$, i.e., at points E and F in *Fig 13*. Hence the roots of $-2x^2+3x+9 = 0$ are $x = -1.5$ and $x = 3$

(d) Comparing $y = -2x^2+3x+6$ (3)
with $0 = -2x^2+x+5$ (4)
shows that if $2x+1$ is added to both sides of equation (4) the right hand side of both equations will be the same. Hence equation (4) may be written as $2x+1 = -2x^2+3x+6$. The solution of this equation is found from the points of intersection of the line $y = 2x+1$ and the parabola $y = -2x^2+3x+6$, i.e., points G and H in *Fig 13*. Hence the roots of $-2x^2+x+5 = 0$ are $x = -1.35$ and $x = 1.85$

*Problem 7* Determine graphically the values of $x$ and $y$ which simultaneously satisfies the equations $y = 2x^2-3x-4$
and $y = 2-4x$

$y = 2x^2-3x-4$ is a parabola and a table of values is drawn up as shown below.

| $x$    | $-2$ | $-1$ | $0$  | $1$  | $2$  | $3$  |
|--------|------|------|------|------|------|------|
| $2x^2$ | 8    | 2    | 0    | 2    | 8    | 18   |
| $-3x$  | 6    | 3    | 0    | $-3$ | $-6$ | $-9$ |
| $-4$   | $-4$ | $-4$ | $-4$ | $-4$ | $-4$ | $-4$ |
| $y$    | 10   | 1    | $-4$ | $-5$ | $-2$ | 5    |

$y = 2-4x$ is a straight line and only three co-ordinates need be calculated.

| $x$ | 0 | 1    | 2    |
|-----|---|------|------|
| $y$ | 2 | $-2$ | $-6$ |

The two graphs are shown plotted in *Fig 14* and the points of intersection, shown as A and B are at co-ordinates $(-2, 10)$ and $(1½, -4)$. Hence the simultaneous solutions occur when $x = -2, y = 10$ and when $x = 1\frac{1}{2}, y = -4$. (These solutions may be checked by substituting into each of the original equations.)

*Problem 8* Solve graphically the cubic equation $4x^3-8x^2-15x+9 = 0$ given that the roots lie between $x = -2$ and $x = 3$. Determine also the co-ordinates of the turning points and distinguish between them.

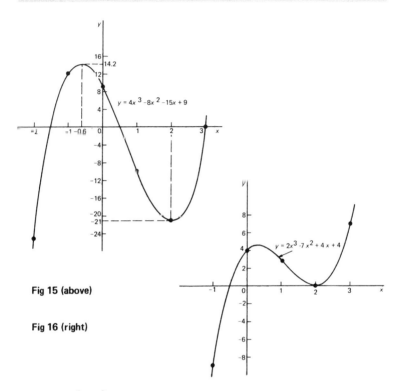

**Fig 15 (above)**

**Fig 16 (right)**

Let $y = 4x^3-8x^2-15x+9$. A table of values is drawn up as shown below.

| $x$ | $-2$ | $-1$ | 0 | 1 | 2 | 3 |
|---|---|---|---|---|---|---|
| $4x^3$ | $-32$ | $-4$ | 0 | 4 | 32 | 108 |
| $-8x^2$ | $-32$ | $-8$ | 0 | $-8$ | $-32$ | $-72$ |
| $-15x$ | 30 | 15 | 0 | $-15$ | $-30$ | $-45$ |
| $+9$ | 9 | 9 | 9 | 9 | 9 | 9 |
| $y$ | $-25$ | 12 | 9 | $-10$ | $-21$ | 0 |

A graph of $y = 4x^3-8x^2-15x+9$ is shown in *Fig 15*. The graph crosses the $x$-axis (where $y = 0$) at $x = -1\frac{1}{2}$, $x = \frac{1}{2}$ and $x = 3$ and these are the solutions to the cubic equation $4x^3-8x^2-15x+9 = 0$. The turning points occur at $(-0.6, 14.2)$, which is a **maximum**, and $(2, -21)$, which is a **minimum**.

**Problem 9** Plot the graph of $y = 2x^3 - 7x^2 + 4x + 4$ for values of $x$ between $x = -1$ and $x = 3$. Hence determine the roots of the equation $2x^3 - 7x^2 + 4x + 4 = 0$.

A table of values is drawn up as shown below.

| $x$    | $-1$ | 0 | 1  | 2   | 3   |
|--------|------|---|----|-----|-----|
| $2x^3$ | $-2$ | 0 | 2  | 16  | 54  |
| $-7x^2$| $-7$ | 0 | $-7$ | $-28$ | $-63$ |
| $+4x$  | $-4$ | 0 | 4  | 8   | 12  |
| $+4$   | 4    | 4 | 4  | 4   | 4   |
| $y$    | $-9$ | 4 | 3  | 0   | 7   |

A graph of $y = 2x^3 - 7x^2 + 4x + 4$ is shown in *Fig 16*. The graph crosses the $x$-axis at $x = -0.5$ and touches the $x$-axis at $x = 2$.
Hence the solutions of the equation $2x^3 - 7x^2 + 4x + 4 = 0$ are $x = -0.5$ and $x = 2$

## C. FURTHER PROBLEMS ON THE GRAPHICAL SOLUTION OF EQUATIONS

In *Problems 1 to 5*, solve the simultaneous equations graphically.
1. $x + y = 2$
   $3y - 2x = 1$            $[x = 1; y = 1]$

2. $y = 5 - x$
   $x - y = 2$            $\left[x = 3\frac{1}{2}; y = 1\frac{1}{2}\right]$

3. $3x + 4y = 5$
   $2x - 5y + 12 = 0$            $[x = -1; y = 2]$

4. $1.4x - 7.06 = 3.2y$
   $2.1x - 6.7y = 12.87$            $[x = 2.3; y = -1.2]$

5. $3x - 2y = 0$
   $4x + y + 11 = 0$            $[x = -2; y = -3]$

6. The friction force $F$ newtons and load $L$ newtons are connected by a law of the form $F = aL + b$, where $a$ and $b$ are constants. When $F = 4$ newtons, $L = 6$ newtons and when $F = 2.4$ newtons, $L = 2$ newtons. Determine graphically the values of $a$ and $b$.            $[a = 0.4; b = 1.6]$

7. Sketch the following graphs and state the nature and co-ordinates of their turning points.
   (a) $y = 4x^2$; (b) $y = 2x^2 - 1$; (c) $y = -x^2 + 3$; (d) $y = -\frac{1}{2}x^2 - 1$
   
   $\left[\begin{array}{l}\text{(a) Minimum } (0, 0); \text{ (b) Minimum } (0, -1); \\ \text{(c) Maximum } (0, 3); \text{ (d) Maximum } (0, -1)\end{array}\right]$

   Solve graphically the quadratic equations in *Problems 8 to 11* by plotting the curves between the given limits. Give answers correct to 1 decimal place.

8. $4x^2 - x - 1 = 0$;    $x = -1$ to $x = 1$            $[-0.4 \text{ or } 0.6]$

9   $x^2 - 3x = 27$;     $x = -5$ to $x = 8$            [$-3.9$ or $6.9$]

10   $2x^2 - 6x - 9 = 0$;     $x = -2$ to $x = 5$       [$-1.1$ or $4.1$]

11   $2x(5x - 2) = 39.6$;     $x = -2$ to $x = 3$.      [$-1.8$ or $2.2$]

12   Solve the quadratic equation $2x^2 + 7x + 6 = 0$ graphically, given that the solutions lie in the range $x = -2$ to $x = 3$. Determine also the nature and co-ordinates of its turning point.

$$\left[x = -1\tfrac{1}{2} \text{ or } 2; \text{ Minimum at } \left(-1\tfrac{3}{4}, -\tfrac{1}{8}\right)\right]$$

13   Solve graphically the quadratic equation $10x^2 - 9x - 11.2 = 0$, given that the roots lie between $x = -1$ and $x = 2$.

$$[x = -0.7 \text{ or } 1.6]$$

14   Plot a graph of $y = 3x^2$ and hence solve the equations (a) $3x^2 - 8 = 0$ and (b) $3x^2 - 2x - 1 = 0$.

$$\left[\text{(a)} \pm 1.63; \text{(b)} 1 \text{ or } -\tfrac{1}{3}\right]$$

15   Plot the graphs $y = 2x^2$ and $y = 3 - 4x$ on the same axes and find the co-ordinates of the points of intersection. Hence determine the roots of the equation $2x^2 + 4x - 3 = 0$.

$$[(-2.58, 13.31), (0.58, 0.67); x = -2.58 \text{ or } 0.58]$$

16   Plot a graph of $y = 10x^2 - 13x - 30$ for values of $x$ between $x = -2$ and $x = 3$. Solve the equation $10x^2 - 13x - 30 = 0$ and from the graph determine (a) the value of $y$ when $x$ is 1.3, (b) the value of $x$ when $y$ is 10 and (c) the roots of the equation $10x^2 - 15x - 18 = 0$.

$$\left[\begin{array}{l} x = -1.2 \text{ or } 2.5 \\ \text{(a)} -30; \text{(b)} 2.75 \text{ and } -1.45; \\ \text{(c)} x = 2.29 \text{ or } -0.79 \end{array}\right]$$

17   Determine graphically the values of $x$ and $y$ which simultaneously satisfy the equations $y = 2(x^2 - 2x - 4)$ and $y + 4 = 3x$.

$$\left[x = 4; y = 8 \text{ and } x = -\tfrac{1}{2}; y = -5\tfrac{1}{2}\right]$$

18   Plot the graph of $y = 4x^2 - 8x - 21$ for values of $x$ from $-2$ to $+4$. Use the graph to find the roots of the following equations: (a) $4x^2 - 8x - 21 = 0$; (b) $4x^2 - 8x - 16 = 0$; (c) $4x^2 - 6x - 18 = 0$.

$$\left[\begin{array}{l}\text{(a)} x = -1.5 \text{ or } 3.5; \text{(b)} x = -1.24 \text{ or } 3.24; \\ \text{(c)} x = -1.5 \text{ or } 3.0 \end{array}\right]$$

19   Plot the graph $y = 4x^3 + 4x^2 - 11x - 6$ between $x = -3$ and $x = 2$ and use the graph to solve the cubic equation $4x^3 + 4x^2 - 11x - 6 = 0$.

$$[x = -2.0; -0.5 \text{ or } 1.5]$$

20   By plotting a graph of $y = x^3 - 2x^2 - 5x + 6$ between $x = -3$ and $x = 4$ solve the equation $x^3 - 2x^2 - 5x + 6 = 0$. Determine also the co-ordinates of the turning points and distinguish between them.

$$(x = -2, 1 \text{ or } 3; \text{ Minimum at } (2.12, -4.10); \text{ Maximum at } (-0.79, 8.21))$$

In *Problems 21 to 24*, solve graphically the cubic equations given, each correct to 2 significant figures.

21   $x^3 - 1 = 0$                                                   [$x = 1$]

22   $x^3 - x^2 - 5x + 2 = 0$              [$x = -2.0, 0.38$ or $2.6$]

23   $x^3 - 2x^2 = 2x - 2$                  [$x = 0.69$ or $2.5$]

24   $2x^3 - x^2 - 9.08x + 8.28 = 0$     [$x = -2.3, 1.0$ or $1.8$]

25   Show that the cubic equation $8x^3 + 36x^2 + 54x + 27$ has only one real root and determine its value.

$$[x = -1.5]$$

# 17 The solution of equations

## A. MAIN POINTS CONCERNED WITH THE SOLUTION OF EQUATIONS

1. An **equation** is a statement that two quantities are equal. To '**solve an equation**' means 'to find the value of the unknown'. The value of the unknown is called the **root** of the equation.
2. (i) A **linear equation** is one in which an unknown quantity is raised to the power 1. For example, $3x+2 = 0$ is a linear equation.
   (ii) A **quadratic equation** is one in which the highest power of the unknown quantity is 2. For example, $2x^2-3x+1 = 0$ is a quadratic equation.
   (iii) A **cubic equation** is one in which the highest power of the unknown quantity is 3. For example, $2x^3+x^2-x+4 = 0$ is a cubic equation. (For the solution of cubic equations, see chapter 16.)
3. There are three methods of solving a pair of **linear simultaneous equations** in two unknowns. These are (i) by substitution; (ii) by elimination; (iii) graphically (see chapter 16). (See *Problems 1 to 7.*)
4. There are four methods of solving **quadratic equations**. These are:
   (i) by factorisation (where possible); (ii) by 'completing the square', (iii) by using the 'quadratic formula'; or (iv) graphically (see chapter 16).
5. Multiplying out $(2x+1)(x-3)$ gives $2x^2-6x+x-3$, i.e. $2x^2-5x-3$. The reverse process of moving from $2x^2-5x-3$ to $(2x+1)(x-3)$ is called **factorising**. If the quadratic expression can be factorised this provides the simplest method of solving a quadratic equation. For example, if $2x^2-5x-3 = 0$, then, by factorising: $(2x+1)(x-3) = 0$.
   Hence either $(2x+1) = 0$, i.e. $x = -\frac{1}{2}$
   or $(x-3) = 0$, i.e. $x = 3$
   The technique of factorising is often one of 'trial and error'. (See *Problems 10 and 11.*)
6. If $ax^2+bx+c = 0$ then $x = \dfrac{-b\pm\sqrt{(b^2-4ac)}}{2a}$
   This is known as the **quadratic formula** and is obtained by '**completing the square**'. (See *Problems 12 to 15.*)
7. There are many **practical problems** where a quadratic equation has first to be obtained, from given information, before it is solved. (See *Problems 16 to 18.*)
8. There are two methods of solving simultaneous equations, one of which is linear and the other quadratic. These are: (i) by substitution (see *Problem 19*) or (ii) graphically (see chapter 16).

167

## B. WORKED PROBLEMS ON THE SOLUTION OF EQUATIONS

*Problem 1* Solve, by a substitution method, the simultaneous equations
$3x-2y = 12$ (1)
$x+3y = -7$ (2)

From equation (2), $x = -7-3y$
Substituting for $x$ in equation (1) gives: $3(-7-3y)-2y = 12$
i.e. $-21-9y-2y = 12$
$-11y = 12+21 = 33$
Hence $y = \dfrac{33}{-11} = -3$
Substituting $y = -3$ in equation (2) gives: $x +3(-3) = -7$
i.e. $x-9 = -7$
Hence $x = -7+9 = 2$

Thus $x = 2$, $y = -3$ is the solution of the simultaneous equations. (Such solutions should always be checked by substituting values into each of the original two equations.)

*Problem 2* Use an elimination method to solve the simultaneous equations
$3x+4y = 5$ (1)
$2x-5y = -12$ (2)

If equation (1) is multiplied throughout by 2 and equation (2) by 3, then the coefficient of $x$ will be the same in the newly formed equations. Thus
2 × equation (1) gives: $6x+ 8y = 10$ (3)
3 × equation (2) gives: $6x-15y = -36$ (4)
Equation (3)−equation (4) gives: $0+23y = 46$
i.e. $y = \dfrac{46}{23} = 2$

(Note $+8y--15y = 8y+15y = 23y$ and $10--36 = 10+36 = 46$. Alternatively, 'change the signs of the bottom line and add').
Substituting $y = 2$ in equation (1) gives: $3x+4(2) = 5$,
from which, $3x = 5-8 = -3$
and $x = -1$
Checking in equation (2), left hand side $= 2(-1)-5(2) = -2-10 = -12 =$ right hand side.
Hence $x = -1$ **and** $y = 2$ is the solution of the simultaneous equations.
The elimination method is the most common method of solving simultaneous equations.

*Problem 3* Solve the equations $\dfrac{x}{2} = 4 - \dfrac{y}{3}$     (1)

$$\dfrac{x}{6} = \dfrac{y}{9} \qquad (2)$$

Removing fractions in equation (1) by multiplying throughout by 6 gives:
$3x = 24 - 2y$
or    $3x + 2y = 24$     (3)
Removing fractions in equation (2) by multiplying throughout by 18 gives:
$3x = 2y$
or    $3x - 2y = 0$     (4)
Adding equations (3) and (4) gives: $6x = 24$, i.e. $x = 4$

Substituting $x = 4$ in equation (1) gives:     $\dfrac{4}{2} = 4 - \dfrac{y}{3}$

from which,    $\dfrac{y}{3} = 4 - 2 = 2$

and    $y = 6$

Checking in equation (2) gives: L.H.S. $= \dfrac{4}{6} = \dfrac{2}{3}$; R.H.S. $= \dfrac{6}{9} = \dfrac{2}{3}$

Hence $x = 4, y = 6$ is the solution of the simultaneous equations.

*Problem 4* Solve the equations $\dfrac{4}{x} - \dfrac{3}{y} = 18$     (1)

$$\dfrac{2}{x} + \dfrac{5}{y} = -4 \qquad (2)$$

The solution of this type of equation is usually easier if the substitution $a = \dfrac{1}{x}$, $b = \dfrac{1}{y}$ is initially made.

Equation (1) becomes:    $4a - 3b = 18$     (3)
Equation (2) becomes:    $2a + 5b = -4$     (4)
2 × equation (4) gives:    $4a + 10b = -8$     (5)

Equation (3) − equation (5) gives:      $-13b = 26$, i.e. $b = \dfrac{26}{-13} = -2$

Substituting $b = -2$ in equation (3) gives:    $4a - 3(-2) = 18$
i.e.    $4a = 18 - 6 = 12$
and    $a = 3$

Checking in equation (4) gives: LHS $= 2(3) + 5(-2) = 6 - 10 = -4 =$ RHS

Since $\dfrac{1}{x} = a$ then $x = \dfrac{1}{a} = \dfrac{1}{3}$

Since $\dfrac{1}{y} = b$ then $y = \dfrac{1}{b} = \dfrac{1}{-2} = -\dfrac{1}{2}$

Hence $x = \dfrac{1}{3}$, $y = -\dfrac{1}{2}$ is the required solution.

*Problem 5* Solve $\dfrac{1}{x+y} = \dfrac{4}{27}$ (1)

$\dfrac{1}{2x-y} = \dfrac{4}{33}$ (2)

To eliminate fractions both sides of equation (1) is multiplied by $27(x+y)$ giving

$27(x+y)\dfrac{1}{x+y} = 27(x+y)\dfrac{4}{27}$

i.e. $27(1) = 4(x+y)$
$27 = 4x+4y$ (3)

Similarly, in equation (2) $33 = 4(2x-y)$
i.e. $33 = 8x-4y$ (4)

Equation (3) + equation (4) gives: $60 = 12x$, i.e. $x = \dfrac{60}{12} = 5$

Substituting $x = 5$ in equation (3) gives: $27 - 4(5)+4y$
from which $4y = 27-20 = 7$

and $y = \dfrac{7}{4} = 1\dfrac{3}{4}$

Hence $x = 5, y = 1\dfrac{3}{4}$ is the required solution, which may be checked in the original equations.

*Problem 6* The molar heat capacity of a solid compound is given by the equation $c = a+bT$, where $a$ and $b$ are constants. When $c = 52$, $T = 100$ and when $c = 172$, $T = 400$. Determine the values of $a$ and $b$.

When $c = 52$, $T = 100$. Hence $52 = a+100b$ (1)
When $c = 172$, $T = 400$. Hence $172 = a+400b$ (2)

Equation (2)−equation (1) gives: $120 = 300b$,

from which $b = \dfrac{120}{300} = 0.4$

Substituting $b = 0.4$ in equation (1) gives: $52 = a+100(0.4)$
$a = 52-40 = 12$

Hence $a = 12$ and $b = 0.4$

*Problem 7* Applying Kirchhoff's laws to an electrical circuit produces the following equations:
$4.5 = I_1 +0.5(I_1 +I_2)$ (1)
$8.5 = 2I_2 +0.5(I_1 +I_2)$ (2)

Determine the values of currents $I_1$ and $I_2$.

Rearranging equation (1) gives: $4.5 = I_1 +0.5I_1 +0.5I_2$
i.e. $1.5I_1 +0.5I_2 = 4.5$ (3)

Rearranging equation (2) gives: $\quad 8.5 = 2I_2 + 0.5I_1 + 0.5I_2$

i.e. $\quad 0.5I_1 + 2.5I_2 = 8.5 \quad$ (4)

3 × equation (4) gives: $\quad 1.5I_1 + 7.5I_2 = 25.5 \quad$ (5)

Equation (5)−equation (3) gives: $\quad 7.0I_2 = 21$

from which $\quad I_2 = \dfrac{21}{7.0} = 3$

Substituting $I_1 = 3$ in equation (3) gives: $\quad 1.5I_1 + 0.5(3) = 4.5$

$\quad 1.5I_1 = 4.5 - 1.5 = 3$

from which $\quad I_1 = \dfrac{3}{1.5} = 2$

Hence $I_1 = 2$ and $I_2 = 3$, which may be checked in the original equations.

*Problem 8* The roots of a quadratic equation are 1/3 and −2. Determine the equation.

If the roots of a quadratic equation are $\alpha$ and $\beta$ then $(x-\alpha)(x-\beta) = 0$

Hence if $\alpha = \dfrac{1}{3}$ and $\beta = -2$, then $(x - \dfrac{1}{3})(x-(-2)) = 0$

i.e. $(x - \dfrac{1}{3})(x+2) = 0$

$x^2 - \dfrac{1}{3}x + 2x - \dfrac{2}{3} = 0$

$x^2 + \dfrac{5}{3}x - \dfrac{2}{3} = 0$

Hence $3x^2 + 5x - 2 = 0$

*Problem 9* Find the equations whose roots are (a) 5 and −5: (b) 1.2 and −0.4

(a) If 5 and −5 are the roots of a quadratic equation then

$(x-5)(x+5) = 0$

i.e. $x^2 - 5x + 5x - 25 = 0$

i.e. $x^2 - 25 = 0$

(b) If 1.2 and −0.4 are the roots of a quadratic equation then

$(x-1.2)(x+0.4) = 0$

i.e. $x^2 - 1.2x + 0.4x - 0.48 = 0$

i.e. $x^2 - 0.8x - 0.48 = 0$

*Problem 10* Solve the equations (a) $x^2 + 2x - 8 = 0$, and (b) $3x^2 - 11x - 4 = 0$ by factorisation.

(a) $x^2 + 2x - 8 = 0$. The factors of $x^2$ are $x$ and $x$. These are placed in brackets thus: $(x \quad )(x \quad )$.

The factors of $-8$ are $+8$ and $-1$, or $-8$ and $+1$, or $+4$ and $-2$, or $-4$ and $+2$.
The only combination to give a middle term of $+2x$ is $+4$ and $-2$.
i.e. $x^2+2x-8 = (x-2)(x+4)$
(Note that the produce of the two inner terms added to the product of the two outer terms must equal the middle term, $+2x$ in this case.)
The quadratic equation $x^2+2x-8 = 0$ thus becomes $(x+4)(x-2) = 0$.
Since the only way that this can be true is for either the first or the second, or both factors to be zero, then
either  $(x+4) = 0$, i.e. $x = -4$
or   $(x-2) = 0$, i.e. $x = 2$
**Hence the roots of $x^2+2x-8 = 0$ are $x = -4$ and 2**
(b) $3x^2 - 11x - 4 = 0$

The factors of $3x^2$ are $3x$ and $x$. These are placed in brackets thus: $(x\ )(3x\ )$.
The factors of $-4$ are $-4$ and $+1$, or $+4$ and $-1$, or $-2$ and $2$. Remembering that the product of the two inner terms added to the product of the two outer terms must equal $-11x$, the only combination to give this is $-4$ and $+1$.
i.e. $3x^2 - 11x - 4 = (3x+1)(x-4)$.
The quadratic equation $3x^2 - 11x - 4 = 0$ thus becomes $(3x+1)(x-4) = 0$.

Hence, either $(3x+1) = 0$, i.e. $x = -\frac{1}{3}$

or $(x-4) = 0$, i.e. $x = 4$
and both solutions may be checked in the original equation.

*Problem 11* Determine the roots of (a) $x^2 - 6x + 9 = 0$, and (b) $4x^2 - 25 = 0$, by factorisation.

(a) $x^2 - 6x + 9 = 0$. Hence $(x-3)(x-3) = 0$
i.e. $(x-3)^2 = 0$, (the left hand side is known as a perfect square).
Hence $x = 3$ is the only root of the equation $x^2 - 6x + 9 = 0$.
(b) $4x^2 - 25 = 0$ (the left hand side is the difference of two squares, $(2x)^2$ and $(5)^2$).
Hence $(2x+5)(2x-5) = 0$.

Hence either $(2x+5) = 0$, i.e. $x = -\frac{5}{2}$

or $(2x-5) = 0$, i.e. $x = \frac{5}{2}$.

*Problem 12* Prove that for the general quadratic equation $ax^2+bx+c = 0$, where $a$, $b$ and $c$ are constants,
$$x = \frac{-b \pm \sqrt{(b^2-4ac)}}{2a}$$

Dividing $ax^2+bx+c = 0$ by $a$ gives: $x^2 + \frac{b}{a}x + \frac{c}{a} = 0$.

Rearranging gives: $x^2 + \frac{b}{a}x = -\frac{c}{a}$

Adding to each side of the equation the square of half the coefficient of the term in $x$ to make the LHS a perfect square gives:

$$x^2 + \frac{b}{a}x + \left(\frac{b}{2a}\right)^2 = \left(\frac{b}{2a}\right)^2 - \frac{c}{a}$$

Rearranging gives: $\left(x + \frac{b}{2a}\right)^2 = \frac{b^2}{4a^2} - \frac{c}{a} = \frac{b^2 - 4ac}{4a^2}$

Taking the square root of both sides gives: $x + \frac{b}{2a} = \sqrt{\left(\frac{b^2-4ac}{4a^2}\right)} = \frac{\pm\sqrt{(b^2-4ac)}}{2a}$

Hence $\qquad x = -\frac{b}{2a} \pm \frac{\sqrt{(b^2-4ac)}}{2a}$

i.e. the quadratic formula is $\qquad x = \frac{-b \pm \sqrt{(b^2-4ac)}}{2a}$

This method of solution is called 'completing the square'.

*Problem 13* Solve $2x^2 + 9x + 8 = 0$, correct to 3 significant figures, by completing the square.

Making the coefficient of $x^2$ unity gives: $x^2 + \frac{9}{2}x + 4 = 0$

and rearranging gives: $\qquad x^2 + \frac{9}{2}x = -4$

Adding to both sides (half the coefficient of $x$)$^2$ gives: $x^2 + \frac{9}{2}x + \left(\frac{9}{4}\right)^2 = \left(\frac{9}{4}\right)^2 - 4$

The LHS is now a perfect square, thus $\left(x + \frac{9}{4}\right)^2 = \frac{81}{16} - 4 = \frac{17}{16}$

Taking the square root of both sides gives: $x + \frac{9}{4} = \sqrt{\left(\frac{17}{16}\right)} = \pm 1.031$

Hence $\qquad x = -\frac{9}{4} \pm 1.031$

i.e. $x = -3.28$ or $-1.22$, correct to 3 significant figures.

*Problem 14* Solve (a) $x^2 + 2x - 8 = 0$, and (b) $3x^2 - 11x - 4 = 0$ by using the quadratic formula.

(a) Comparing $x^2 + 2x - 8 = 0$ with $ax^2 + bx + c = 0$ gives $a = 1$, $b = 2$ and $c = -8$.

Substituting these values into the quadratic formula $x = \frac{-b \pm \sqrt{(b^2-4ac)}}{2a}$

gives: $x = \frac{-2 \pm \sqrt{[(2)^2 - 4(1)(-8)]}}{2(1)} = \frac{-2 \pm \sqrt{(4+32)}}{2} = \frac{-2 \pm \sqrt{36}}{2}$

$= \frac{-2 \pm 6}{2} = \frac{-2+6}{2}$ or $\frac{-2-6}{2}$

Hence $x = \frac{4}{2} = 2$ or $\frac{-8}{2} = -4$ (as in *Problem 10(a)*).

(b) Comparing $3x^2 - 11x - 4 = 0$ with $ax^2 + bx + c = 0$ gives $a = 3$, $b = -11$ and $c = -4$.

Hence $x = \dfrac{-(-11) \pm \sqrt{[(-11)^2 - 4(3)(-4)]}}{2(3)} = \dfrac{+11 \pm \sqrt{(121 + 48)}}{6} = \dfrac{11 \pm \sqrt{169}}{6}$

$= \dfrac{11 \pm 13}{6} = \dfrac{11 + 13}{6}$ or $\dfrac{11 - 13}{6}$

Hence $x = \dfrac{24}{6} = 4$ or $\dfrac{-2}{6} = -\dfrac{1}{3}$ (as in *Problem 10(b)*).

*Problem 15* Solve $4x^2 + 7x + 2 = 0$ giving the roots correct to 2 decimal places.

Comparing $4x^2 + 7x + 2 = 0$ with $ax^2 + bx + c = 0$ gives $a = 4$, $b = 7$ and $c = 2$.

Hence $x = \dfrac{-7 \pm \sqrt{[(7)^2 - 4(4)(2)]}}{2(4)} = \dfrac{-7 \pm \sqrt{17}}{8} = \dfrac{-7 \pm 4.123}{8} = \dfrac{-7 + 4.123}{8}$ or $\dfrac{-7 - 4.123}{8}$

Hence $x = -0.36$ or $-1.39$, correct to 2 decimal places.

*Problem 16* The area of a rectangle is 23.6 cm² and its width if 3.10 cm shorter than its length. Determine the dimensions of the rectangle, correct to 3 significant figures.

Let the length of the rectangle be $x$ cm. Then the width is $(x - 3.10)$ cm.
Area = length × width = $x(x - 3.10) = 23.6$
i.e. $x^2 - 3.10x - 23.6 = 0$

Using the quadratic formula, $x = \dfrac{-(-3.10) \pm \sqrt{[(-3.10)^2 - 4(1)(-23.6)]}}{2(1)}$

$= \dfrac{3.10 \pm \sqrt{(9.61 + 94.4)}}{2} = \dfrac{3.10 \pm 10.20}{2}$

$= \dfrac{13.30}{2}$ or $\dfrac{-7.10}{2}$

Hence $x = 6.65$ cm or $-3.55$ cm. The latter solution is neglected since length cannot be negative. Thus length $x = 6.65$ cm and width $= x - 3.10 = 6.65 - 3.10 = 3.55$ cm.

**Hence the dimensions of the rectangle are 6.61 cm by 3.55 cm.**
(Check: Area = 6.65 × 3.55 = 23.6 cm², correct to 3 significant figures.)

*Problem 17* Calculate the diameter of a solid cylinder which has a height of 82.0 cm and a total surface area of 2.0 m².

Total surface area of a cylinder = curved surface area + 2 circular ends
$= 2\pi rh + 2\pi r^2$ (where $r$ = radius and $h$ = height)

Since the total surface area = 2.0 m² and the height $h$ = 82 cm or 0.82 m
then $2.0 = 2\pi r(0.82) + 2\pi r^2$
i.e. $2\pi r^2 + 2\pi r(0.82) - 2.0 = 0$

Dividing throughout by $2\pi$ gives: $r^2 + 0.82r - \dfrac{1}{\pi} = 0$

Using the quadratic formula: $r = \dfrac{-0.82 \pm \sqrt{[(0.82)^2 - 4(1)(-\frac{1}{\pi})]}}{2(1)}$

$= \dfrac{-0.82 \pm \sqrt{1.9456}}{2} = \dfrac{-0.82 \pm 1.3948}{2}$

$= 0.2874$ or $-1.1074$.

Thus the radius $r$ of the cylinder is 0.2874 m (the negative solution being neglected).
Hence the diameter of the cylinder = 2 × 0.2874
= 0.5748 m or 57.5 cm, correct to 3 significant figures.

---

*Problem 18* The height $s$ metres of a mass projected vertically upwards at time $t$ seconds is $s = ut - \frac{1}{2}gt^2$. Determine how long the mass will take after being projected to reach a height of 16 m (a) on the ascent and (b) on the descent, when $u$ = 30 m/s and $g$ = 9.81 m/s².

When height $s$ = 16 m, $16 = 30t - \dfrac{1}{2}(9.81)t^2$
i.e. $4.905t^2 - 30t + 16 = 0$

Using the quadratic formula: $t = \dfrac{-(-30) \pm \sqrt{[(-30)^2 - 4(4.905)(16)]}}{2(4.905)}$

$= \dfrac{30 \pm \sqrt{586.1}}{9.81} = \dfrac{30 \pm 24.21}{9.81} = 5.53$ or $0.59$

**Hence the mass will reach a height of 16 m after 0.59 s on the ascent and after 5.53 s on the descent.**

---

*Problem 19* Determine the values of $x$ and $y$ which simultaneously satisfy the equations $y = 2x^2 - 3x - 4$ and $y = 2 - 4x$.

For a simultaneous solution the values of $y$ must be equal. Hence, equating the two equations gives: $2x^2 - 3x - 4 = 2 - 4x$
i.e. $2x^2 - 3x - 4 - 2 + 4x = 0$
or $2x^2 + x - 6 = 0$

Factorising gives: $(2x - 3)(x + 2) = 0$

from which, $x = \dfrac{3}{2}$ or $-2$

In the equation $y = 2 - 4x$, when $x = \dfrac{3}{2}$, $y = 2 - 4\left(\dfrac{3}{2}\right) = -4$

and when $x = -2$, $y = 2 - 4(-2) = +10$

Hence the simultaneous solutions occur when $x = 1\frac{1}{2}$, $y = -4$ and when $x = -2$, $y = 10$ (which may be checked in $y = 2x^2 - 3x - 4$).
A graphical solution of this problem is shown in *Problem 7,* chapter 16, page 163.

## C. FURTHER PROBLEMS ON THE SOLUTION OF EQUATIONS

In *Problems 1 to 8,* solve the given simultaneous equations by the elimination or substitution method and verify the results.

1. $3x + 2y = 12$
   $4x - y = 5$
   $$[x = 2; y = 3]$$

2. $5x - 3y = 11$
   $3x + y = 8$
   $$\left[x = 2\frac{1}{2}; y = \frac{1}{2}\right]$$

3. $5x = 2y$
   $3x + 11 = 7y$
   $$[x = 2; y = 5]$$

4. $\frac{x}{2} - 7 + 2y = 0$
   $5x + \frac{2}{3}y = 12$
   $$[x = 2; y = 3]$$

5. $3y - 2.5x = 0.45$
   $1.6x + 0.8y = 0.8$
   $$[x = 0.30; y = 0.40]$$

6. $\frac{4}{x} - \frac{3}{y} = 18$
   $\frac{2}{x} + \frac{5}{y} = -4$
   $$[x = \frac{1}{3}; y = -\frac{1}{2}]$$

7. $\frac{x+1}{4} + 1 - \frac{y+2}{3} = 0$
   $\frac{1-x}{5} + \frac{3-y}{4} + \frac{13}{20} = 0$
   $$[x = 3; y = 4]$$

8. $\frac{x}{4} + \frac{2+y}{3} = \frac{1}{2}$
   $\frac{5+x}{6} + \frac{2y}{5} = \frac{9}{10}$
   $$[x = -2; y = 1]$$

9. In a simple machine, the effort $E$ newtons required to overcome a load $L$ newtons is given by: $E = a + bL$, where $a$ and $b$ are constants. An effort of 4.2 N overcomes a load of 5 N and an effort of 5 N overcomes a load of 6 N. Determine the values of $a$ and $b$.
   $$[a = 0.2, b = 0.8]$$

10. Equations produced by applying Kirchhoff's laws to an electrical circuit are:
    $4 = 0.3I_1 + 1.5(I_1 - I_2)$
    $10 = 1.5(I_2 - I_1) + 0.2I_2$.
    Determine currents $I_1$ and $I_2$, correct to 2 decimal places.
    $$[I_1 = 26.91; I_2 = 29.63]$$

11  The distance $s$ travelled by a vehicle in $t$ seconds is given by:
$s = ut + \frac{1}{2}at^2$, where $u$ and $a$ are constants. Determine the values of $u$ and $a$ given that $s = 34$ m when $t = 2$s, and $s = 167.5$ m when $t = 5$s.

$[u = 6; a = 11]$

12  Determine the quadratic equations in $x$ whose roots are (a) 3 and 1, (b) 2 and $-5$, (c) $-1$ and $-4$.

$[(a)\ x^2 - 4x + 3 = 0;\ (b)\ x^2 + 3x - 10 = 0;\ (c)\ x^2 + 5x + 4 = 0]$

13  Determine the quadratic equations in $x$ whose roots are (a) $2\frac{1}{2}$ and $-\frac{1}{2}$; (b) 6 and $-6$, (c) 2.4 and $-0.7$.

$[(a)\ 4x^2 - 8x - 5 = 0;\ (b)\ x^2 - 36 = 0;\ (c)\ x^2 - 1.7x - 1.68 = 0]$

In *Problems 14 to 16*, solve the given equations by factorisation.

14  (a) $x^2 + 4x - 32 = 0$
    (b) $x^2 - 16 = 0$
    (c) $(x+2)^2 = 16$

$\begin{bmatrix} (a)\ 4, -8 \\ (b)\ 4, -4 \\ (c)\ 2, -6 \end{bmatrix}$

15  (a) $2x^2 - x - 3 = 0$
    (b) $6x^2 - 5x + 1 = 0$
    (c) $10x^2 + 3x - 4 = 0$

$\begin{bmatrix} (a)\ -1, 1\frac{1}{2} \\ (b)\ \frac{1}{2}, \frac{1}{3} \\ (c)\ \frac{1}{2}, -\frac{4}{5} \end{bmatrix}$

16  (a) $x^2 - 4x + 4 = 0$
    (b) $21x^2 - 25x = 4$
    (c) $8x^2 + 13x - 6 = 0$

$\begin{bmatrix} (a)\ 2 \\ (b)\ 1\frac{1}{3}, -\frac{1}{7} \\ (c)\ \frac{3}{8}, -2 \end{bmatrix}$

In *Problems 17 to 19*, solve the given equations by using the quadratic formula or by completing the square, correct to 3 decimal places.

17  (a) $x^2 + 4x + 1 = 0$
    (b) $2x^2 + 5x - 4 = 0$
    (c) $3x^2 - x - 5 = 0$

$\begin{bmatrix} (a)\ -3.732, -0.268 \\ (b)\ -3.137, 0.637 \\ (c)\ 1.468, -1.135 \end{bmatrix}$

18  (a) $5x^2 - 8x + 2 = 0$
    (b) $4x^2 - 11x + 3 = 0$
    (c) $2x^2 + 5x = 2$

$\begin{bmatrix} (a)\ 1.290, 0.310 \\ (b)\ 2.443, 0.307 \\ (c)\ -2.851, 0.351 \end{bmatrix}$

19  (a) $4x + 5 = \dfrac{3}{x}$
    (b) $(2x+1) = \dfrac{5}{x-3}$
    (c) $\dfrac{x+1}{x-1} = x - 3$

$\begin{bmatrix} (a)\ 0.443, -1.693 \\ (b)\ 3.608, -1.108 \\ (c)\ 4.562, 0.438 \end{bmatrix}$

In *Problems 20 to 22* determine the solutions of the simultaneous equations.

20  $y = x^2 + x + 1$
    $y = 4 - x$

$[x = 1, y = 3 \text{ and } x = -3, y = 7]$

21  $y = 15x^2 + 21x - 11$
    $y = 2x - 1$

$\left[x = \dfrac{2}{5}, y = -\dfrac{1}{5} \text{ and } x = -1\dfrac{2}{3}, y = -4\dfrac{1}{3}\right]$

22  $2x^2 + y = 4 + 5x$
    $x + y = 4$

$[x = 0, y = 4 \text{ and } x = 3, y = 1]$

23. The angle a rotating shaft turns through in $t$ seconds is given by $\theta = \omega t + \frac{1}{2}\alpha t^2$. Determine the time taken to complete 4 radians if $\omega$ is 3.0 rad/s and $\alpha$ is 0.60 rad/s$^2$.

[1.191 s]

24. The power $P$ developed in an electrical circuit is given by $P = 10I - 8I^2$, where $I$ is the current in amperes. Determine the current necessary to produce a power of 2.5 watts in the circuit.

[0.345 A or 0.905 A]

25. The area of a triangle is 47.6 cm$^2$ and its perpendicular height is 4.3 cm more than its base length. Determine the length of the base correct to 3 significant figures.

[7.84 cm]

26. The sag $l$ metres in a cable stretched between two supports, distance $x$ m apart is given by:

$$l = \frac{12}{x} + x.$$

Determine the distance between supports when the sag is 20 m.

[0.619 m or 19.38 m]

27. The acid dissociation constant $K_a$ of ethanoic acid is $1.8 \times 10^{-5}$ mol dm$^{-3}$ for a particular solution. Using the Ostwald dilution law

$$K_a = \frac{x^2}{v(1-x)}$$

determine $x$, the degree of ionisation, given that $v = 10$ dm$^3$.

[0.0133]

28. A rectangular building is 15 m long by 11 m wide. A concrete path of constant width is laid all the way around the building. If the area of the path is 60.0 m$^2$, calculate its width correct to the nearest millimetre.

[1.066 m]

29. The total surface area of a closed cylindrical container is 20.0 m$^3$. Calculate the radius of the cylinder if its height is 2.80 m.

[86.78 cm]

30. The bending moment $M$ at a point in a beam is given by

$$M = \frac{3x(20-x)}{2},$$

where $x$ metres is the distance from the point of support. Determine the value of $x$ when the bending moment is 50 N m.

[1.835 m or 18.165 m]

# 18 Exponential functions and Naperian logarithms

## A. MAIN POINTS CONCERNED WITH EXPONENTIAL FUNCTIONS AND NAPERIAN LOGARITHMS

1. An **exponential function** is one which contains $e^x$, $e$ being a constant called the **exponent** and having an approximate value of 2.7183. The exponent arises from the natural laws of growth and decay and is used as a base of natural or Naperian logarithms.

2. **The natural laws of growth and decay** are of the form $y = Ae^{kx}$, where $A$ and $k$ are constants. The natural laws occur frequently in engineering and science and examples of quantities related by a natural law include:
   - (i) Linear expansion $\qquad\qquad\qquad\qquad\qquad\; l = l_0 e^{\alpha\theta}$
   - (ii) Change in electrical resistance with temperature $\quad R_\theta = R_0 e^{\alpha\theta}$
   - (iii) Tension in belts $\qquad\qquad\qquad\qquad\qquad\; T_1 = T_0 e^{\mu\alpha}$
   - (iv) Newton's law of cooling $\qquad\qquad\qquad\qquad \theta = \theta_0 e^{-kt}$
   - (v) Biological growth $\qquad\qquad\qquad\qquad\qquad y = y_0 e^{kt}$
   - (vi) Discharge of a capacitor $\qquad\qquad\qquad\qquad q = Qe^{-t/CR}$
   - (vii) Atmospheric pressure $\qquad\qquad\qquad\qquad\; p = p_0 e^{-h/c}$
   - (viii) Radioactive decay $\qquad\qquad\qquad\qquad\qquad N = N_0 e^{-\lambda t}$
   - (ix) Decay of current in an inductive circuit $\qquad\quad i = Ie^{-Rt/L}$

3. The **value of $e^x$** may be determined by using
   - (i) a calculator which possesses an '$e^x$' function,
   - (ii) the power series $e^x = 1 + x + \dfrac{x^2}{2!} + \dfrac{x^3}{3!} + \ldots\ldots$ (where 3! is 'factorial 3' and means $3 \times 2 \times 1$),
   - (iii) Naperian logarithms (see para. 7), or
   - (iv) 4 figure tables of exponential functions which enable values of $e^x$ and $e^{-x}$ to be read over a range of $x$ from 0.02 to 6.0
   
   For example, $e^{0.36} = 1.4333$, $e^{4.3} = 73.700$, $e^{-0.47} = 0.6250$ and $e^{-2.6} = 0.0743$.
   Also, from tables and laws of indices,
   $e^{0.68} = e^{0.6 + 0.08} = (e^{0.6})(e^{0.08}) = (1.8221)(1.0833)$
   $\qquad\; = 1.9739$ correct to 4 decimal places.
   
   (See *Problems 1 and 2*.)

4. **Graphs of exponential functions**
   - (i) Values of $e^x$ and $e^{-x}$, obtained from 4 figure tables, correct to 2 decimal places, over the range $x = -3$ to $x = 3$, are shown in the table below.

| $x$ | -3.0 | -2.5 | -2.0 | -1.5 | -1.0 | -0.5 | 0 | 0.5 | 1.0 | 1.5 | 2.0 | 2.5 | 3.0 |
|---|---|---|---|---|---|---|---|---|---|---|---|---|---|
| $e^x$ | 0.05 | 0.08 | 0.14 | 0.22 | 0.37 | 0.61 | 1.00 | 1.65 | 2.72 | 4.48 | 7.39 | 12.18 | 20.09 |
| $e^{-x}$ | 20.09 | 12.18 | 7.39 | 4.48 | 2.72 | 1.65 | 1.00 | 0.61 | 0.37 | 0.22 | 0.14 | 0.08 | 0.05 |

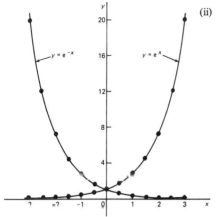

**Fig 1**

*Fig 1* shows graphs of $y = e^x$ and $y = e^{-x}$.

(ii) A similar table may be drawn up for $y = 5e^{\frac{1}{2}x}$, a graph of which is shown in *Fig 2*. The gradient of the curve at any point, $dy/dx$, is obtained by drawing a tangent to the curve at that point and measuring the gradient of the tangent. For example:

when $x = 0$, $y = 5$ and $\dfrac{dy}{dx} = \dfrac{BC}{AB}$

$= \dfrac{(6.2 - 3.7)}{1} = 2.5$

and when $x = 2$, $y = 13.6$ and

$\dfrac{dy}{dx} = \dfrac{EF}{DE} = \dfrac{(16.8 - 10)}{1} = 6.8$

These two results each show that $\dfrac{dy}{dx} = \dfrac{1}{2}y$, and further determinations of the gradients of $y = 5e^{\frac{1}{2}x}$ would give the same result for each. In general, for all natural growth and decay laws of the form $y = Ae^{kx}$, where $k$ is a positive constant for growth laws (as in *Fig 2*) and a negative constant for decay curves, $dy/dx = ky$, i.e., **the rate of change of the variable, $y$, is proportional to the variable itself.**

(iii) For any natural law of growth and decay of the form $dy/dx = ky$, the solution is always $y = Ae^{kx}$. (see *Problems 3 to 9*).

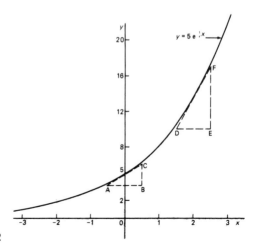

**Fig 2**

**Hyperbolic or Naperian logarithms**

5  (i) A **logarithm** of a number is the power to which a base has to be raised to be equal to the number. Thus, if $y = a^x$ then $x = \log_a y$.

  (ii) Logarithms having a base of 10 are called **common logarithms** and the common logarithm of $x$ is written as $\lg x$.

  (iii) Logarithms having a base of $e$ are called **hyperbolic, Naperian** or **natural logarithms** and the Naperian logarithm of $x$ is written as $\log_e x$, or more commonly, $\ln x$.

6  The **value of a Naperian logarithm** may be obtained by using:

  (i) a calculator possessing a '$\ln x$' function,

  (ii) the change of base rule for logarithms, which states, $\log_a y = \dfrac{\log_b y}{\log_b a}$,

  from which, $\ln y = \dfrac{\lg y}{\lg e} = \dfrac{\lg y}{0.4343} = 2.3026 \lg y$,

or (iii) 4 figure Naperian logarithm tables.

7  **Use of 4 figure Naperian logarithm tables**

  (i) **For numbers from 1 to 10**, Naperian logarithm tables are used in a similar way to common logarithms. For example,
   $\ln 3.4 = 1.2238$
   $\ln 3.47 = 1.2442$
   $\ln 3.478 = 1.2442 + 23$ (from the mean difference column equivalent to 8)
   i.e. $\ln 3.478 = 1.2465$
   Similarly, $\ln 5.731 = 1.7459$ and $\ln 9.159 = 2.2148$

  (ii) **Numbers larger than 10** are initially expressed in standard form and the supplementary table of Naperian logarithms of $10^{+n}$ used. For example,
   $\ln 64 = \ln(6.4 \times 10^1) = \ln 6.4 + \ln 10^1$
   $= 1.8563 + 2.3026 = 4.1589$
   Similarly, $\ln 327.6 = \ln(3.276 \times 10^2) = \ln 3.276 + \ln 10^2$
   $= 1.1866 + 4.6052 = 5.7918$.

  (iii) **Numbers smaller than 1** are initially expressed in standard form and the supplementary table of Naperian logarithms of $10^{-n}$ used.
   For example, $\ln 0.064 = \ln(6.4 \times 10^{-2}) = \ln 6.4 + \ln 10^{-2}$
   $= 1.8563 + \bar{5}.3948 = \bar{3}.2511$
   $\bar{3}.2511$ is the same as $-3 + 0.2511 = -2.7489$ (which is the value given by a calculator).
   Similarly, $\ln 0.003276 = \ln 3.276 \times 10^{-3}) = \ln 3.276 + \ln 10^{-3}$
   $= 1.1866 + \bar{7}.0922 = \bar{6}.2788$ or $-5.7212$.

  (iv) **The antilogarithm of a number between 0 and 2.3026** is obtained by finding the value of the logarithm within the table, the antilogarithm being obtained from the corresponding row and column. For example, if $\ln x = 1.4317$, the nearest number less than this is 1.4303 corresponding to 4.18. The difference between 1.4317 and 1.4303 is 14, which in the mean difference column corresponds to 6. Hence if $\ln x = 1.4317$ then $x = 4.186$.

  (v) **To find the antilogarithm of a number greater than 2.3026**, for example, 5.3726, the nearest number less than 5.3726 is found in the supplementary table of Naperian logarithms of $10^{+n}$; in this case it is 4.6052, corresponding to $10^2$. The difference between 5.3726 and 4.6052 is 0.7674. The antilogarithm of 0.7674 is read directly from within the table as 2.154. Hence if $\ln x = 5.3726$ then $x = 2.154 \times 10^2 = 215.4$.

(vi) To find the antilogarithm of a number less than 2.3026, for example, $-5.8314$, the number is initially changed to $\bar{6}.1686$ and the nearest more negative number in the supplementary table of Naperian logarithms of $10^{-n}$ is found; in this case it is $\bar{7}.0922$, corresponding to $10^{-3}$. The difference between $\bar{6}.1686$ and $\bar{7}.0922$ is $1.0764$. The antilogarithm of $1.0764$ is read directly from within the tables as $2.934$. Hence if $\ln x = -5.8314$ then $x = 2.934 \times 10^{-3} = 0.002\ 934$.

(vii) The procedure for determining antilogarithms is considerably quicker with a **calculator**. If $\ln x = y$ then $x = e^y$, from the definition of a logarithm. Thus, if $\ln x = 5.3726$, then $x = e^{5.3726} = 215.42224$, by calculator and if $\ln x = -5.8314$ then $x = e^{-5.8314} = 2.9339 \times 10^{-3} = 0.002\ 933\ 9$, by calculator. Alternatively, an 'invert $\ln x$' function is available on many calculators.

(See *Problems 10 to 16*.)

## B WORKED PROBLEMS ON EXPONENTIAL FUNCTIONS AND NAPERIAN LOGARITHMS

*Problem 1* Use exponential tables to determine the values of (a) $e^{0.18}$, (b) $e^{-1.4}$, (c) $3e^{-0.54}$, (d) $0.2e^8$, each correct to 4 significant figures.

From exponential tables:
(a) $e^{0.18} = 1.1972 = \mathbf{1.197}$, correct to 4 significant figures.
(b) $e^{-1.4} = \mathbf{0.2466}$
(c) $e^{-0.54} = (e^{-0.5})(e^{-0.04}) = (0.6065)(0.9608) = 0.582\ 73$,
   Hence $3e^{-0.54} = 3(0.582\ 73) = \mathbf{1.748}$, correct to 4 significant figures.
(d) $e^8 = e^{4+4} = (e^4)(e^4) = (54.598)^2 = 2980.94$.
   Hence $0.2e^8 = (0.2)(2980.94) = \mathbf{596.2}$, correct to 4 significant figures.

*Problem 2* Determine the value of $5e^{0.5}$, correct to 5 significant figures by using the power series for $e^x$.

$$e^x = 1 + x + \frac{x^2}{2!} + \frac{x^3}{3!} + \ldots\ldots\ldots\ldots$$

Hence $e^{0.5} = 1 + 0.5 + \dfrac{(0.5)^2}{(2)(1)} + \dfrac{(0.5)^3}{(3)(2)(1)} + \dfrac{(0.5)^4}{(4)(3)(2)(1)} + \dfrac{(0.5)^5}{(5)(4)(3)(2)(1)}$
$\qquad\qquad\qquad + \dfrac{(0.5)^6}{(6)(5)(4)(3)(2)(1)}$

$\qquad\quad = 1 + 0.5 + 0.125 + 0.020\ 833 + 0.002\ 604\ 2 + 0.000\ 260\ 4 + 0.000\ 021\ 7$
i.e. $e^{0.5}\ = 1.648\ 72$, correct to 6 significant figures.
Hence $5e^{0.5} = 5(1.64872) = \mathbf{8.2436}$, correct to 5 significant figures.

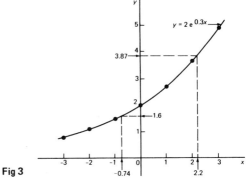

**Fig 3**

*Problem 3* Plot a graph of $y = 2e^{0.3x}$ over a range of $x = -3$ to $x = 3$. Hence determine the value of $y$ when $x = 2.2$ and the value of $x$ when $y = 1.6$.

A table of values is drawn up as shown below.

| $x$ | $-3$ | $-2$ | $-1$ | 0 | 1 | 2 | 3 |
|---|---|---|---|---|---|---|---|
| $0.3x$ | $-0.9$ | $-0.6$ | $-0.3$ | 0 | 0.3 | 0.6 | 0.9 |
| $e^{0.3x}$ | 0.407 | 0.549 | 0.741 | 1.00 | 1.350 | 1.822 | 2.460 |
| $2e^{0.3x}$ | 0.81 | 1.10 | 1.48 | 2.00 | 2.70 | 3.64 | 4.92 |

A graph of $y = 2e^{0.3x}$ is shown plotted in *Fig 3*.
When $x = 2.2$, $y = 3.87$ and when $y = 1.6$, $x = -0.74$

*Problem 4* Plot the graph of $y = \frac{1}{3}e^{-2x}$ over the range $x = -1.5$ to $x = 1.5$. Determine, from the graph, the value of $y$ when $x = -1.2$ and the value of $x$ when $y = 1.4$

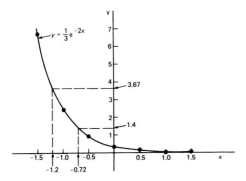

**Fig 4**

A table of values is drawn up as shown below.

| $x$ | −1.5 | −1.0 | −0.5 | 0 | 0.5 | 1.0 | 1.5 |
|---|---|---|---|---|---|---|---|
| $-2x$ | 3 | 2 | 1 | 0 | −1 | −2 | −3 |
| $e^{-2x}$ | 20.086 | 7.389 | 2.718 | 1.00 | 0.368 | 0.135 | 0.050 |
| $\frac{1}{3}e^{-2x}$ | 6.70 | 2.46 | 0.91 | 0.33 | 0.12 | 0.05 | 0.02 |

A graph of $y = \frac{1}{3}e^{-2x}$ is shown in *Fig 4*

When $x = -1.2$, $y = \mathbf{3.67}$ and when $y = 1.4$, $x = \mathbf{-0.72}$

*Problem 5* A natural law of growth is of the form $y = 4e^{0.2x}$. Plot a graph depicting this law for values of $x$ from $x = -3$ to $x = 3$. From the graph determine (a) the value of $y$ when $x$ is 2.2, (b) the value of $x$ when $y$ is 3.4 and (c) the rate of change of $y$ with respect to $x$ (i.e. $dy/dx$) at $x = -2$

A table of values is drawn up as shown below.

| $x$ | −3 | −2 | −1 | 0 | 1 | 2 | 3 |
|---|---|---|---|---|---|---|---|
| $0.2x$ | −0.6 | −0.4 | −0.2 | 0 | 0.2 | 0.4 | 0.6 |
| $e^{0.2x}$ | 0.549 | 0.670 | 0.819 | 1.00 | 1.221 | 1.492 | 1.822 |
| $4e^{0.2x}$ | 2.20 | 2.68 | 3.28 | 4.00 | 4.88 | 5.97 | 7.29 |

A graph of $y = 4e^{0.2x}$ is shown in *Fig 5*. From the graph:
(a) $x = 2.2$, $y = \mathbf{6.2}$,
(b) when $y = 3.4$, $x = \mathbf{-0.8}$,
(c) at $x = -2$, gradient

(i.e. $\frac{dy}{dx}$) $= \frac{BC}{AB} = \frac{1.08}{2} = \mathbf{0.54}$

[From para. 4, when $y = Ae^{kx}$
then $dy/dx = ky$
In this case $A = 4$, $k = 0.2$ thus
$y = 4e^{0.2x}$.
Hence, when $x = -2$,
$dy/dx = (0.2)4e^{0.2(-2)} = 0.54$]

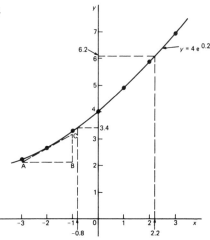

**Fig 5**

**Problem 6** The decay of voltage, $V$ volts, across a capacitor at time $t$ seconds is given by $v = 250e^{-t/3}$. Draw a graph showing the natural decay curve over the first 6 s. From the graph find (a) the voltage after 3.4 s, and (b) the time when the voltage is 150 V. (c) Determine the rate of change of voltage after 2 s and after 4 s.

A table of values is drawn up as shown below.

| $t$ | 0 | 1 | 2 | 3 | 4 | 5 | 6 |
|---|---|---|---|---|---|---|---|
| $e^{-t/3}$ | 1.00 | 0.7165 | 0.5134 | 0.3679 | 0.2636 | 0.1889 | 0.1353 |
| $v = 250e^{-t/3}$ | 250.0 | 179.1 | 128.4 | 91.97 | 65.90 | 47.22 | 33.83 |

The natural decay curve of $v = 250e^{-t/3}$ is shown in *Fig 6*.

From the graph:
(a) when time $t = 3.4$ s, voltage $v = 80$ volts, and
(b) when voltage $v = 150$ volts, time $t = 1.5$ s.
(c) Rate of change of voltage (i.e. $\frac{dv}{dt}$) after 2 s is given by $\frac{BC}{AB}$.

$$\frac{BC}{AB} = \frac{(170-85)}{(1-3)} = \frac{-85}{2} = -42.5 \text{ V/s}.$$

After 4 s, $\frac{dv}{dt} = \frac{EF}{DE} = \frac{(77-55)}{(3.5-4.5)}$
$= \frac{-22}{1} = -22$ V/s.

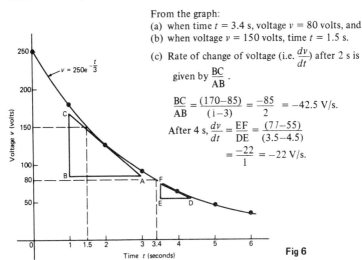

**Fig 6**

**Problem 7** A law of natural growth is of the form $dy/dx = 8y$. Determine the solution of this equation.

An equation of the form $dy/dx = ky$ has the solution $y = Ae^{kx}$, where $A$ and $k$ are constants (see para. 4). With the equation $dy/dx = 8y$, $k = 8$ thus the solution is $y = Ae^{8x}$.

**Problem 8** The rate of change of charge on a capacitor $dQ/dt$ is proportional to the charge $Q$ and is given by $R\frac{dQ}{dt} + \frac{Q}{C} = 0$, where $R$ and $C$ are constants. Solve the equation for $Q$.

Rearranging $R\frac{dQ}{dt} + \frac{Q}{C} = 0$ gives $R\frac{dQ}{dt} = -\frac{Q}{C}$

Thus $\frac{dQ}{dt} = -\frac{Q}{CR}$, i.e. $\frac{dQ}{dt} = \left(-\frac{1}{CR}\right) Q$ which is of the form $\frac{dy}{dx} = ky$,

where $k = -\frac{1}{CR}$.

The solution is $Q = Ae^{(-\frac{1}{CR})t} = Ae^{-\frac{t}{CR}}$.

***Problem 9*** **The rate of decay of radioactive atoms, $dN/dt$, is proportional to the number of radioactive atoms $N$ and is given by $dN/dt + 0.8 \times 10^4 N = 0$. Solve the equation for $N$.**

Rearranging $\frac{dN}{dt} + 0.8 \times 10^4 N = 0$ gives $\frac{dN}{dt} = -0.8 \times 10^4 N$, which is of the form $\frac{dy}{dx} = ky$, where $k = -0.8 \times 10^4$.

Hence the solution is $N = Ae^{-0.8 \times 10^4 t}$

***Problem 10*** **Use tables to determine (a) ln 7.483; (b) ln 748.3; (c) ln 0.007 483.**

(a) $\ln 7.4 = 2.0015$
$\ln 7.48 = 2.0122$
$\ln 7.483 = 2.0122 + 4$ (from the mean difference column equivalent to 3)
i.e. $\ln 7.483 = \mathbf{2.0126}$

(b) From para. 7(ii), $\ln 748.3 = \ln(7.483 \times 10^2) = \ln 7.483 + \ln 10^{-2}$
$= 2.0126 + 4.6052 = \mathbf{6.6178}$

(c) From para. 7(iii), $\ln 0.007\,483 = \ln(7.483 \times 10^{-3}) = \ln 7.483 + \ln 10^{-3}$
$= 2.0126 + \bar{7}.0922 = \mathbf{\bar{5}.1048}$ **or** $\mathbf{-4.8952}$

***Problem 11*** **Determine from tables the value of $x$ when $\ln x$ is (a) 1.7321; (b) 8.4628; (c) –8.4217.**

(a) Since 1.7321 lies between 0 and 2.3026 the value of $x$ can be read directly from within the tables of Naperian logarithms. The nearest number less than 1.7321 is 1.7317 corresponding to 5.65. The difference between 1.7321 and 1.7317 is 4 which corresponds to 2 in the mean difference column.
Hence if $\ln x = 1.7321$, then $x = \mathbf{5.652}$.
(Alternatively, if $\ln x = 1.7321$, then $x = e^{1.7321} = 5.652\,511\,7$ by calculator– or an 'invert $\ln x$' is available on many calculators.)

(b) The nearest number less than 8.4628 in the supplementary table of Naperian logarithms of $10^{+n}$ is 6.9078 corresponding to $10^3$. The difference between

8.4628 and 6.9078 is 1.5550. The antilogarithm of 1.5550 is found from within the tables and is 4.735.
Hence if $\ln x = 8.4628$, then $x = 4.735 \times 10^3 =$ **4735**.
(Alternatively, $x = e^{8.4628} = 4735.2983$ by calculator.)

(c) $-8.4217 = -9 + 0.5783 = \overline{9}.5783$

The nearest more negative number to $\overline{9}.5783$ in the supplementary table of Naperian logarithms of $10^{-n}$ is $\overline{10}.7897$, corresponding to $10^{-4}$. The difference between $\overline{9}.5783$ and $\overline{10}.7897$ is 0.7886. The antilogarithm of 0.7886 is found from within the tables and is 2.200.
Hence if $\ln x = -8.4217$, then $x =$ **2.200 $\times 10^{-4}$**.

*Problem 12* Use common logarithms to evaluate $\ln 36.25$.

From the change of base rule, $\ln y = 2.3026 \lg y$ (from para. 6)
Hence $\ln 36.25 = (2.3026)(\lg 36.25)$
$= (2.3026)(1.5593) =$ **3.5904**, correct to 4 decimal places,

which may be checked by using Naperian logarithms or a calculator.

*Problem 13* Use Naperian logarithms to evaluate (a) $e^{4.137}$ and (b) $-4.5e^{-2.4379}$, each correct to 3 significant figures.

(a) Let $y = e^{4.137}$. Taking Naperian logarithms of each side gives:
$\ln y = \ln e^{4.137} = 4.137 \ln e$ by the laws of logarithms.
Thus $\ln y = 4.137$, since $\ln e = 1$.
$\ln y = 2.3026 + (4.137 - 2.3026)$
$= 2.3026 + 1.8344$.
Taking antilogarithms of both sides gives $y = 10^1 \times 6.261$
i.e. $e^{4.137} =$ **62.6**, correct to 3 significant figures.

(b) Let $y = e^{-2.4379}$. Taking Naperian logarithms of each side gives:
$\ln y = -2.4379 = \overline{3}.5621 = \overline{5}.3948 + (\overline{3}.5621 - \overline{5}.3948)$
i.e. $\ln y = \overline{5}.3948 + 2.1673$
Taking antilogarithms of both sides gives $y = 10^{-2} \times 8.734$.
Thus $-4.5e^{-2.4379} = (-4.5)(8.734 \times 10^{-2}) =$ **−0.393**, correct to 3 significant figures.

*Problem 14* The resistance $R$ of an electrical conductor at temperature $\theta°C$ is given by $R = R_0 e^{\alpha\theta}$, where $\alpha$ is a constant and $R_0 = 5 \times 10^3$ ohms. Determine the value of $\alpha$ when $R = 6 \times 10^3$ ohms and $\theta = 1500°C$. Also, find the temperature when the resistance $R$ is $5.4 \times 10^3$ ohms.

Transposing $R = R_0 e^{\alpha\theta}$ gives $\dfrac{R}{R_0} = e^{\alpha\theta}$

Taking Naperian logarithms of both sides gives: $\ln\left(\dfrac{R}{R_0}\right) = \ln e^{\alpha\theta} = \alpha\theta$

Hence $\alpha = \frac{1}{\theta} \ln\left(\frac{R}{R_0}\right) = \frac{1}{1500} \ln\left(\frac{6 \times 10^3}{5 \times 10^3}\right) = \frac{1}{1500} (0.1823)$

Hence $\alpha = 1.215 \times 10^{-4}$ (by calculator or tables).

From above, $\ln\left(\frac{R}{R_0}\right) = \alpha\theta$, hence $\theta = \frac{1}{\alpha} \ln\left(\frac{R}{R_0}\right)$

When $R = 5.4 \times 10^3$, $\alpha = 1.215 \times 10^{-4}$ and $R_0 = 5 \times 10^3$,

$\theta = \frac{1}{1.215 \times 10^{-4}} \ln\left(\frac{5.4 \times 10^3}{5 \times 10^3}\right) = \frac{10^4}{1.215} (7.696 \times 10^{-2}) = \mathbf{633.4°C}$.

***

**Problem 15** In an experiment involving Newton's law of cooling, the temperature $\theta$ (°C) is given by $\theta = \theta_0 e^{-kt}$. Find the value of constant $k$ when $\theta_0 = 56.6°C$, $\theta = 16.5°C$ and $t = 83.0$ seconds.

***

Transposing $\theta = \theta_0 e^{-kt}$ gives $\frac{\theta}{\theta_0} = e^{-kt}$ from which $\frac{\theta_0}{\theta} = \frac{1}{e^{-kt}} = e^{kt}$.

Taking Naperian logarithms of both sides gives: $\ln\left(\frac{\theta_0}{\theta}\right) = kt$,

from which, $k = \frac{1}{t} \ln\left(\frac{\theta_0}{\theta}\right) = \frac{1}{83.0} \ln\left(\frac{56.6}{16.5}\right) = \frac{1}{83.0} (1.2326)$

Hence $k = \mathbf{1.485 \times 10^{-2}}$

***

**Problem 16** The current $i$ amperes flowing in a capacitor at time $t$ seconds is given by $i = 8.0(1 - e^{-t/CR})$, where the circuit resistance $R$ is $25 \times 10^3$ ohms and capacitance $C$ is $16 \times 10^{-6}$ farads. Determine (a) the current $i$ after 0.5 seconds and (b) the time for the current to reach 6.0 A.

***

(a) Current $i = 8.0(1 - e^{-t/CR}) = 8.0\left[1 - e^{-\frac{0.5}{(16 \times 10^{-6})(25 \times 10^3)}}\right] = 8.0(1 - e^{-1.25})$

$= 8.0(1 - 0.2865) = 8.0(0.7135) = \mathbf{5.708\ amperes}$.

(b) Transposing $i = 8.0(1 - e^{-t/CR})$ gives $\frac{i}{8.0} = 1 - e^{-t/CR}$,

from which, $e^{-t/CR} = 1 - \frac{i}{8.0} = \frac{8.0 - i}{8.0}$

Taking the reciprocal of both sides gives: $e^{t/CR} = \frac{8.0}{8.0 - i}$

Taking Naperian logarithms of both sides gives: $\frac{t}{CR} = \ln\left(\frac{8.0}{8.0 - i}\right)$

Hence $t = CR \ln\left(\frac{8.0}{8.0 - i}\right) = (16 \times 10^{-6})(25 \times 10^3) \ln\left(\frac{8.0}{8.0 - 6.0}\right)$

when $i = 6.0$ amperes,

i.e. $t = \frac{400}{10^3} \ln\left(\frac{8.0}{2.0}\right) = 0.4 \ln 4.0 = 0.4(1.3863) = \mathbf{0.5545\ s}$

## C. FURTHER PROBLEMS ON EXPONENTIAL FUNCTIONS AND NAPERIAN LOGARITHMS

In *Problems 1 and 2* use exponential tables to evaluate the given functions correct to 4 significant figures. Check your results by using a calculator.

1  (a) $e^{4.4}$; (b) $e^{-0.25}$; (c) $e^{0.92}$
$$[(a)\ 81.45;\ (b)\ 0.7788;\ (c)\ 2.509]$$

2  (a) $e^{-1.8}$; (b) $e^{-0.78}$; (c) $e^{10}$
$$[(a)\ 0.1653;\ (b)\ 0.4584;\ (c)\ 22\,030]$$

3  Use the power series for $e^x$ to determine, correct to 4 significant figures, (a) $e^2$; (b) $e^{-0.3}$ and check your result by using exponential tables and/or a calculator.
$$[(a)\ 7.389;\ (b)\ 0.7408]$$

4  Evaluate, correct to 4 significant figures, (a) $3.5e^{2.8}$; (b) $-\frac{6}{5}e^{-1.5}$; (c) $2.16e^{5.7}$
$$[(a)\ 57.56;\ (b)\ -0.2678;\ (c)\ 645.6]$$

5  Plot a graph of $y = 3e^{0.2x}$ over the range $x = -3$ to $x = 3$. Hence determine the value of $y$ when $x = 1.4$ and the value of $x$ when $y = 4.5$.
$$[3.97;\ 2.03]$$

6  Plot a graph of $y = \frac{1}{2}e^{-1.5x}$ over a range $x = -1.5$ to $x = 1.5$ and hence determine the value of $y$ when $x = -0.8$ and the value of $x$ when $y = 3.5$.
$$[1.66;\ -1.30]$$

7  Plot a graph of $y = 2.5e^{-0.15x}$ over a range $x = -8$ to $x = 8$. Determine from the graph the value of $y$ when $x = -6.2$ and the value of $x$ when $y = 5.4$.
$$[6.34;\ -5.13]$$

8  Draw a graph of $y = 2(2e^{-x} - 3e^{2x})$ over a range of $x = -3$ to $x = 3$. Determine the value of $y$ when $x = -2.2$ and the value of $x$ when $y = 17.4$.
$$[36.0;\ -1.49]$$

9  In a chemical reaction the amount of starting material $C$ cm$^3$ left after $t$ minutes is given by $C = 40e^{-0.006t}$. Plot a graph of $C$ against $t$ and determine (a) the concentration $C$ after 1 hour; (b) the time taken for the concentration to decrease by half; (c) Determine the rate of change of $C$ with $t$ after 40 mins.
$$[(a)\ 27.9\ \text{cm}^3;\ (b)\ 115.5\ \text{mins};\ (c)\ -0.189\ \text{cm}^3/\text{min}]$$

10  The rate at which a body cools is given by $\theta = 250e^{-0.05t}$, where the excess of temperature of a body above its surroundings at time $t$ minutes is $\theta\,°C$. Plot a graph showing this natural decay curve for the first hour of cooling and hence determine the rate of cooling after (a) 15 minutes; (b) 45 minutes.
$$[(a)\ -5.90°C/\text{min};\ (b)\ -1.32°C/\text{min}]$$

11  The tensions in two sides of a belt, $T$ and $T_0$ newtons, passing round a pulley wheel and in contact with the pulley for an angle $\theta$ radians is given by $T = T_0 e^{0.3\theta}$. Plot a graph depicting this relationship over a range $\theta = 0$ to $\theta = 2.0$ radians, given $T_0 = 50$ N. From the graph determine the value of $dT/d\theta$ when $\theta = 1.2$ radians.
$$[21.5\ \text{N/rad}]$$

12  The voltage drop, $v$ volts, across an inductor is related to time, $t$ ms, by $v = 30 \times 10^3 e^{-t/10}$. Plot a graph of $v$ against $t$ from $t = 0$ to $t = 10$ ms. Use the graph to determine the rate of change of voltage with time (i.e. $dv/dt$) when $t = 5.5$ ms.
$$[-1731\ \text{V/ms}]$$

13  Determine the solution of the following equations:
(a) $\dfrac{dT}{d\theta} = 0.4T$; (b) $\dfrac{dm}{di} - 2m = 0$; (c) $\dfrac{dv}{dx} + 9.81v = 0$; (d) $\dfrac{1}{5}\dfrac{dm}{d\theta} - \dfrac{1}{2}m = 0$
$$\left[(a)\ T = Ae^{0.4\theta};\ (b)\ m = Ae^{2i};\ (c)\ v = Ae^{-9.81x};\ (d)\ m = Ae^{\frac{5}{2}\theta}\right]$$

14 The change of length $l$ of a bar of metal with respect to temperature $\theta$ is directly proportional to its length and may be represented by $(dl/d\theta) - \alpha l = 0$, where $\alpha$ is a constant equal to $2.5 \times 10^{-6}$. Solve the differential equation for $l$.

$$[l = Ae^{2.5 \times 10^{-6} \theta}]$$

15 The rate of change of voltage across an electrical circuit, $dV/dt$, is directly proportional to the applied voltage $V$ such that $7.5V - 5.0(dV/dt) = 0$. Solve the equation for $V$.

$$[V = Ae^{1.5t}]$$

In *Problems 16 to 18* use 4 figure tables to evaluate the given functions.

16 (a) ln 1.73; (b) ln 5.413; (c) ln 9.412

[(a) 0.5481; (b) 1.6887; (c) 2.2420]

17 (a) ln 17.3; (b) ln 541.3; (c) 9412

[(a) 2.8507; (b) 6.2939; (c) 9.1498]

18 (a) ln 0.173; (b) ln 0.005 413; (c) ln 0.094 12

[(a) $\bar{2}$.2455 or $-1.7545$; (b) $\bar{6}$.7809 or $-5.2191$; (c) $\bar{3}$.6368 or $-2.3632$]

In *Problems 19 to 21* use 4 figure tables to find the value of $x$ when ln $x$ has the value shown.

19 (a) 0.4317; (b) 1.2047; (c) 2.0491

[(a) 1.540; (b) 3.336; (c) 7.761]

20 (a) 3.1429; (b) 6.3312; (c) 10.3171

[(a) 23.17; (b) 561.8; (c) 30 250]

21 (a) $\bar{1}$.3617; (b) $\bar{6}$.3173; (c) $-3.3117$

[(a) 0.5282; (b) 0.003 404; (c) 0.036 45]

22 Use common logarithms to evaluate (a) ln 29.41 and (b) ln 110.5, each correct to 4 significant figures.

[(a) 3.381; (b) 4.705]

23 Use Naperian logarithms to evaluate, correct to 3 significant figures,

(a) $e^{3.912}$; (b) $2.7e^{-1.8463}$; (c) $-\dfrac{15}{8} e^{2.3131}$

[(a) 50.0; (b) 0.426; (c) $-18.9$]

24 Evaluate $0.26e^{-\frac{3x}{7}}$ when $x$ has a value of (a) $-3.68$; (b) 2.417; (c) 14.

[(a) 1.2587; (b) 0.092 28; (c) $6.445 \times 10^{-4}$]

25 Two quantities $x$ and $y$ are related by the equation $y = ae^{-kx}$, where $a$ and $k$ are constants. (a) Determine the value of $y$ when $a = 2.114$, $k = -3.20$ and $x = 1.429$. (b) Determine the value of $x$ when $y = 115.4$, $a = 17.8$ and $k = 4.65$.

[(a) 204.7; (b) $-0.4020$]

26 The pressure $p$ pascals at height $h$ metres above ground level is given by $p = p_0 e^{-h/C}$, where $p_0$ is the pressure at ground level and $C$ is a constant. When $p_0$ is $1.012 \times 10^5$ Pa and the pressure at a height of 1420 m is $9.921 \times 10^4$ Pa, determine the value of $C$.

[71 500]

27 The length $l$ metres of a metal bar at temperature $t°C$ is given by $l = l_0 e^{\alpha t}$, where $l_0$ and $\alpha$ are constants. Determine (a) the value of $\alpha$ when $l = 1.993$ m, $l_0 = 1.894$ m and $t = 250°C$, and (b) the value of $l_0$ when $l = 2.416$, $t = 310°C$ and $\alpha = 1.682 \times 10^{-4}$.

[(a) $2.038 \times 10^{-4}$; (b) 2.293 m]

28 The temperature $\theta_2 °C$ of an electrical conductor at time $t$ seconds is given by $\theta_2 = \theta_1(1 - e^{-t/T})$, where $\theta_1$ is the initial temperature and $T$ seconds is a constant. Determine (a) $\theta_1$ when $\theta_2 = 50°C$, $t = 30$ s and $T = 80$ s, and (b) the time $t$ for $\theta_2$ to fall to half the value of $\theta_1$ if $T$ remains at 80 s.

[(a) 159.9°C; (b) 55.45 s]

29 Quantities $x$ and $y$ are related by $y = 8.317(1-e^{cx/t})$, where $c$ and $t$ are constants. Determine (a) the value of $y$ when $c = 2.9 \times 10^{-3}$, $x = 841.2$ and $t = 4.379$, and (b) the value of $t$ when $y = -83.68$, $x = 841.2$ and $c = 2.9 \times 10^{-2}$.

[(a) −6.201; (b) 10.15]

30 The voltage drop, $v$ volts, across an inductor $L$ henrys at time $t$ seconds is given by $v = 200e^{-Rt/L}$, where $R = 150 \,\Omega$ and $L = 12.5 \times 10^{-3}$ H. Determine (a) the voltage when $t = 160 \times 10^{-6}$ s and (b) the time for the voltage to reach 85 V.

[(a) 29.32 volts; (b) $71.31 \times 10^{-6}$ s]

# 19 Boolean algebra and switching circuits

## A. MAIN POINTS CONCERNED WITH BOOLEAN ALGEBRA AND SWITCHING CIRCUITS

1. A **two-state device** is one whose basic elements can only have one of two conditions. Thus, two-way switches, which can either be on or off, and the binary numbering system, having the digits 0 and 1 only, are two-state devices. In Boolean algebra, if $A$ represents one state, then $\bar{A}$, called 'not-$A$', represents the second state.

2. The **or**-function
   In Boolean algebra, the **or**-function for two elements $A$ and $B$ is written as $A+B$, and is defined as '$A$, or $B$, or both $A$ and $B$'. The equivalent electrical circuit for a two-input **or**-function is given by two switches connected in parallel. With reference to *Fig 1(a)*, the lamp will be on when $A$ is on, when $B$ is on, or when both $A$ and $B$ are on. In the table shown in *Fig 1(b)*, all the possible switch combinations are shown in columns 1 and 2, in which a 0 represents a switch being off and a 1 represents the switch being on, these columns being called the inputs. Column 3 is called the output and a 0 represents the lamp being off and a 1 represents the lamp being on. Such a table is called a **truth table**.

3. The **and**-function
   In Boolean algebra, the **and**-function for two elements $A$ and $B$ is written as $A.B$ and is defined as 'both $A$ and $B$'. The equivalent electrical circuit for a two-input **and**-function is given by two switches connected in series. With reference to *Fig 2(a)* the lamp will be on only when both $A$ and $B$ are on. The truth table for a two-input **and**-function is shown in *Fig 2(b)*.

4. The **not**-function
   In Boolean algebra, the **not**-function for element $A$ is written as $\bar{A}$, and is defined as 'the opposite to $A$'. Thus if $A$ means switch $A$ is on, $\bar{A}$ means that switch $A$ is off. The truth table for the **not**-function is shown in *Table 1*.

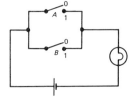

| 1 | 2 | 3 |
|---|---|---|
| Input (switches) | | Output (lamp) |
| A | B | Z = A + B |
| 0 | 0 | 0 |
| 0 | 1 | 1 |
| 1 | 0 | 1 |
| 1 | 1 | 1 |

**Fig 1**

(a) Switching circuit for **or** – function   (b) Truth table for **or** – function

5 In paras 2, 3 and 4 above, the Boolean expressions, equivalent switching circuits and truth tables for the three functions used in Boolean algebra are given for a two-input system. A system may have more than two inputs and the Boolean expression for a three-input **or**-function having elements $A$, $B$ and $C$ is $A+B+C$. Similarly, a three-input **and**-function is written as $A.B.C$. The equivalent electrical circuits and truth tables for three-input **or** and **and**-functions are shown in *Figs 3(a) and (b)* respectively.

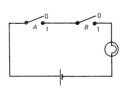

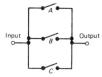

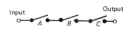

TABLE 1

| *Input* $A$ | *Output* $Z = \overline{A}$ |
|---|---|
| 0 | 1 |
| 1 | 0 |

(a) Switching circuit for **and** – function

| Input (switches) $A$ | $B$ | Output (lamp) $Z = A.B$ |
|---|---|---|
| 0 | 0 | 0 |
| 0 | 1 | 0 |
| 1 | 0 | 0 |
| 1 | 1 | 1 |

(b) Truth table for **and** – function

**Fig 2**

| Input $A$ | $B$ | $C$ | Output $Z = A + B + C$ |
|---|---|---|---|
| 0 | 0 | 0 | 0 |
| 0 | 0 | 1 | 1 |
| 0 | 1 | 0 | 1 |
| 0 | 1 | 1 | 1 |
| 1 | 0 | 0 | 1 |
| 1 | 0 | 1 | 1 |
| 1 | 1 | 0 | 1 |
| 1 | 1 | 1 | 1 |

(a) The **or** – function electrical circuit and truth table

| Input $A$ | $B$ | $C$ | Output $Z = A.B.C$ |
|---|---|---|---|
| 0 | 0 | 0 | 0 |
| 0 | 0 | 1 | 0 |
| 0 | 1 | 0 | 0 |
| 0 | 1 | 1 | 0 |
| 1 | 0 | 0 | 0 |
| 1 | 0 | 1 | 0 |
| 1 | 1 | 0 | 0 |
| 1 | 1 | 1 | 1 |

(b) The **and** – function electrical circuit and truth table

**Fig 3**

| 1 | 2 | 3 | 4 | 5 |
|---|---|---|---|---|
| A | B | $A.B$ | $\bar{A}.\bar{B}$ | $Z = AB + \bar{A}.\bar{B}$ |
| 0 | 0 | 0 | 1 | 1 |
| 0 | 1 | 0 | 0 | 0 |
| 1 | 0 | 0 | 0 | 0 |
| 1 | 1 | 1 | 0 | 1 |

(a) Truth table for $Z = A.B + \bar{A}.\bar{B}$

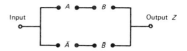

(b) Switching circuit for $Z = A.B + \bar{A}.\bar{B}$.

**Fig 4**

TABLE 2 Some Laws and Rules of Boolean Algebra

| Reference | Name | Rule or law |
|---|---|---|
| 1 | Commutative laws | $A+B = B+A$ |
| 2 | | $A.B = B.A$ |
| 3 | Associative laws | $(A+B)+C = A+(B+C)$ |
| 4 | | $(A.B).C = A.(B.C)$ |
| 5 | Distributive laws | $A.(B+C) = A.B+A.C$ |
| 6 | | $A+(B.C) = (A+B).(A+C)$ |
| 7 | | $A+0 = A$ |
| 8 | Sum | $A+1 = 1$ |
| 9 | rules | $A+A = A$ |
| 10 | | $A+\bar{A} = 1$ |
| 11 | | $A.0 = 0$ |
| 12 | Product | $A.1 = A$ |
| 13 | rules | $A.A = A$ |
| 14 | | $A.\bar{A} = 0$ |

6  To achieve a given output, it is often necessary to use combinations of switches connected both in series and in parallel. If the output from a switching circuit is given by the Boolean expression $Z = A.B+\bar{A}.\bar{B}$, the truth table is as shown in *Fig 4(a)*. In this table, columns 1 and 2 give all the possible combinations of $A$ and $B$. Column 3 corresponds to $A.B$ and column 4 to $\bar{A}.\bar{B}$, i.e. a 1 output is obtained when $A = 0$ and when $B = 0$. Column 5 is the **or**-function applied to columns 3 and 4 giving an output of $Z = A.B+\bar{A}.\bar{B}$. The corresponding switching circuit is shown in *Fig 4(b)* in which $A$ and $B$ are connected in series to give $A.B$, $\bar{A}$ and $\bar{B}$ are connected in series to give $\bar{A}.\bar{B}$, and $A.B$ and $\bar{A}.\bar{B}$ are connected in parallel to give $A.B+\bar{A}.\bar{B}$. The circuit symbols used are such that $A$ means the switch is on when $A$ is 1, $\bar{A}$ means the switch is on when $A$ is 0, and so on.

7 When describing a complex switching circuit by means of a Boolean expression, often many terms and many elements per term are used. Frequently, the laws and rules of Boolean algebra may be used to simplify the Boolean expression and some of the laws and rules are given in *Table 2*. These rules and laws may be verified by using a truth table, as shown in *Problems 5 and 6*.

## B. WORKED PROBLEMS ON BOOLEAN ALGEBRA AND SWITCHING CIRCUITS

*Problem 1* Derive the Boolean expression and construct a truth table for the switching circuit shown in *Fig 5(a)*.

The switches between 1 and 2 in *Fig 5(a)* are in series and have a Boolean expression of $B.A$. The parallel circuit 1 to 2 and 3 to 4 have a Boolean expression of $(B.A+\bar{B})$. The parallel circuit can be treated as a single switching unit, giving the

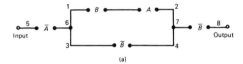

(a)

| 1<br>A | 2<br>B | 3<br>B.A | 4<br>$\bar{B}$ | 5<br>$B.A + \bar{B}$ | 6<br>$\bar{A}$ | 7<br>$Z = \bar{A}.(B.A + \bar{B}).\bar{B}$ |
|---|---|---|---|---|---|---|
| 0 | 0 | 0 | 1 | 1 | 1 | 1 |
| 0 | 1 | 0 | 0 | 0 | 1 | 0 |
| 1 | 0 | 0 | 1 | 1 | 0 | 0 |
| 1 | 1 | 1 | 0 | 1 | 0 | 0 |

(b)

**Fig 5**

equivalent of switches 5 to 6, 6 to 7 and 7 to 8 in series. Thus the output is given by

$Z = \bar{A}.(B.A+\bar{B}).\bar{B}$

The truth table is as shown in *Fig 5(b)*. Columns 1 and 2 give all the possible combinations of switches $A$ and $B$. Column 3 is the **and**-function applied to columns 1 and 2, giving $B.A$. Column 4 is $\bar{B}$, i.e., the opposite to column 2. Column 5 is the **or**-function applied to columns 3 and 4. Column 6 is $\bar{A}$, i.e., the opposite to column 1. The output is column 7 and is obtained by applying the **and**-function to columns 4, 5 and 6.

*Problem 2* Derive the Boolean expression and construct a truth table for the switching circuit shown in *Fig 6(a)*.

The parallel circuit 1 to 2 and 3 to 4 gives $(A+\bar{B})$ and this is equivalent to a single switching unit between 7 and 2. The parallel circuit 5 to 6 and 7 to 2 gives $C+(A+\bar{B})$ and this is equivalent to a single switching unit between 8 and 2. The series circuit 9 to 8 and 8 to 2 gives the output $Z = B.[C+(A+\bar{B})]$.

The truth table is shown in *Fig 6(b)*. Columns 1, 2 and 3 give all the possible combinations of $A$, $B$ and $C$. Column 4 is $\bar{B}$ and is the opposite to column 2.

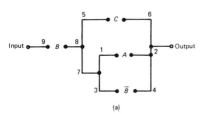

Column 5 is the **or**-function applied to columns 1 and 4, giving $(A+\bar{B})$. Column 6 is the **or**-function applied to columns 3 and 5 giving $C+(A+\bar{B})$. The output is given in column 7 and is obtained by applying the **and**-function to columns 2 and 6, giving $Z = B.[C+(A+\bar{B})]$.

(a)

| 1 | 2 | 3 | 4 | 5 | 6 | 7 |
|---|---|---|---|---|---|---|
| A | B | C | $\bar{B}$ | $A+\bar{B}$ | $C+(A+\bar{B})$ | $Z = B.[C+(A+\bar{B})]$ |
| 0 | 0 | 0 | 1 | 1 | 1 | 0 |
| 0 | 0 | 1 | 1 | 1 | 1 | 0 |
| 0 | 1 | 0 | 0 | 0 | 0 | 0 |
| 0 | 1 | 1 | 0 | 0 | 1 | 1 |
| 1 | 0 | 0 | 1 | 1 | 1 | 0 |
| 1 | 0 | 1 | 1 | 1 | 1 | 0 |
| 1 | 1 | 0 | 0 | 1 | 1 | 1 |
| 1 | 1 | 1 | 0 | 1 | 1 | 1 |

(b)

**Fig 6**

*Problem 3* Construct a switching circuit to meet the requirements of the Boolean expression:

$Z = A.\bar{C}+\bar{A}.B+\bar{A}.B.\bar{C}$

Construct the truth table for this circuit.

The three terms joined by **or**-functions, (+), indicate three parallel branches, having:

        branch 1    $A$ and $\bar{C}$ in series
        branch 2    $\bar{A}$ and $B$ in series
and  branch 3    $\bar{A}$ and $B$ and $\bar{C}$ in series

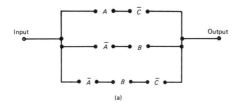

| 1<br>A | 2<br>B | 3<br>C | 4<br>$\bar{C}$ | 5<br>$A.\bar{C}$ | 6<br>$\bar{A}$ | 7<br>$\bar{A}.B$ | 8<br>$\bar{A}.B.\bar{C}$ | 9<br>$Z = A.\bar{C} + \bar{A}.B + \bar{A}.B.\bar{C}$ |
|---|---|---|---|---|---|---|---|---|
| 0 | 0 | 0 | 1 | 0 | 1 | 0 | 0 | 0 |
| 0 | 0 | 1 | 0 | 0 | 1 | 0 | 0 | 0 |
| 0 | 1 | 0 | 1 | 0 | 1 | 1 | 1 | 1 |
| 0 | 1 | 1 | 0 | 0 | 1 | 1 | 0 | 1 |
| 1 | 0 | 0 | 1 | 1 | 0 | 0 | 0 | 1 |
| 1 | 0 | 1 | 0 | 0 | 0 | 0 | 0 | 0 |
| 1 | 1 | 0 | 1 | 1 | 0 | 0 | 0 | 1 |
| 1 | 1 | 1 | 0 | 0 | 0 | 0 | 0 | 0 |

(b)

**Fig 7**

Hence the required switching circuit is as shown in *Fig 7(a)*.
The corresponding truth table is shown in *Fig 7(b)*.
Column 4 is $\bar{C}$, i.e., opposite to column 3.
Column 5 is $A.\bar{C}$, obtained by applying the **and**-function to columns 1 and 4.
Column 6 is $\bar{A}$, the opposite to column 1.
Column 7 is $\bar{A}.B$, obtained by applying the **and**-function to columns 2 and 6.
Column 8 is $\bar{A}.B.\bar{C}$, obtained by applying the **and**-function to columns 4 and 7.
Column 9 is the output, obtained by applying the **or**-function to columns 5, 7 and 8.

*Problem 4* Derive the Boolean expression and construct the switching circuit for the truth table given in *Fig 8(a)*.

Examination of the truth table shown in *Fig 8(a)* shows that there is a 1 output in the Z-column in rows 1, 3, 4 and 6. Thus, the Boolean expression and switching circuit should be such that a 1 output is obtained for row 1 **or** row 3 **or** row 4 **or** row 6. In row 1, $A$ is 0 **and** $B$ is 0 **and** $C$ is 0 and this corresponds to the Boolean expression $\bar{A}.\bar{B}.\bar{C}$. In row 3, $A$ is 0 **and** $B$ is 1 **and** $C$ is 0, i.e., the Boolean expression is $\bar{A}.B.\bar{C}$. Similarly in rows 4 and 6, the Boolean expressions are $\bar{A}.B.C$ and $A.\bar{B}.C$ respectively. Hence the Boolean expression is:

$Z = \bar{A}.\bar{B}.\bar{C} + \bar{A}.B.\bar{C} + \bar{A}.B.C + A.\bar{B}.C$

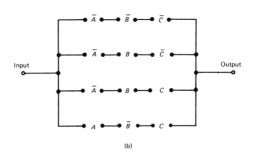

### Fig 8

The corresponding switching circuit is shown in *Fig 8(b)*. The four terms are joined by **or**-functions (+), and are represented by four parallel circuits. Each term has three elements joined by an **and**-function, and is represented by three elements connected in series.

*Problem 5* By constructing the appropriate truth table show that
$A+(B.C) = (A+B).(A+C)$

The truth table is shown in *Table 3(a)*.

TABLE 3(a)

| 1 | 2 | 3 | 4 | 5 | 6 | 7 | 8 |
|---|---|---|---|---|---|---|---|
| $A$ | $B$ | $C$ | $B.C$ | $(A+B.C)$ | $A+B$ | $A+C$ | $(A+B).(A+C)$ |
| 0 | 0 | 0 | 0 | 0 | 0 | 0 | 0 |
| 0 | 0 | 1 | 0 | 0 | 0 | 1 | 0 |
| 0 | 1 | 0 | 0 | 0 | 1 | 0 | 0 |
| 0 | 1 | 1 | 1 | 1 | 1 | 1 | 1 |
| 1 | 0 | 0 | 0 | 1 | 1 | 1 | 1 |
| 1 | 0 | 1 | 0 | 1 | 1 | 1 | 1 |
| 1 | 1 | 0 | 0 | 1 | 1 | 1 | 1 |
| 1 | 1 | 1 | 1 | 1 | 1 | 1 | 1 |

TABLE 3(b)

| 1 | 2 | 3 | 4 | 5 | 6 |
|---|---|---|---|---|---|
| $A$ | $B$ | $\overline{A}$ | $\overline{A}.B$ | $A+\overline{A}.B$ | $A+B$ |
| 0 | 0 | 1 | 0 | 0 | 0 |
| 0 | 1 | 1 | 1 | 1 | 1 |
| 1 | 0 | 0 | 0 | 1 | 1 |
| 1 | 1 | 0 | 0 | 1 | 1 |

Columns 1, 2 and 3 are all the possible combinations of $A$, $B$ and $C$.
Column 4 is $B.C$, obtained by applying the **and**-function to columns 2 and 3.
Column 5 is $(A+B.C)$, obtained by applying the **or**-function to columns 1 and 4.
Column 6 is $(A+B)$, obtained by applying the **or**-function to columns 1 and 2.
Column 7 is $(A+C)$, obtained by applying the **or**-function to columns 1 and 3.
Column 8 is $(A+B).(A+C)$, obtained by applying the **and**-function to columns 6 and 7.
Since the pattern of 0's and 1's in columns 5 and 8 are the same, this shows that
$A+(B.C) = (A+B).(A+C)$.

*Problem 6* Show that $A+\overline{A}.B = A+B$

In the truth table shown in *Table 3(b)*, since columns 5 and 6 have the same pattern of 0's and 1's, then

$A+\overline{A}.B = A+B$

*Problem 7* Simplify the Boolean expression:

$\overline{P}.\overline{Q}+\overline{P}.Q+P.\overline{Q}$

The laws and rules of Boolean algebra given in *Table 2* are used to simplify Boolean expressions.

|  | *Reference* |
|---|---|
| $\overline{P}.\overline{Q}+\overline{P}.Q+P.\overline{Q} = \overline{P}.(\overline{Q}+Q)+P.\overline{Q}$ | 5 |
| $= \overline{P}.1+P.\overline{Q}$ | 10 |
| $= \overline{P}+P.\overline{Q}$ | 12 |

*Problem 8* Simplify $(P+\overline{P}.Q).(Q+\overline{Q}.P)$

With reference to *Table 2*:

|  | *Reference* |
|---|---|
| $(P+\overline{P}.Q).(Q+\overline{Q}.P) = P.(Q+\overline{Q}.P)+\overline{P}.Q.(Q+\overline{Q}.P)$ | 5 |
| $= P.Q+P.\overline{Q}.P+\overline{P}.Q.Q+\overline{P}.Q.\overline{Q}.P$ | 5 |
| $= P.Q+P.\overline{Q}+\overline{P}.Q+\overline{P}.Q.\overline{Q}.P$ | 13 |
| $= P.Q+P.\overline{Q}+\overline{P}.Q+0$ | 14 |
| $= P.Q+P.\overline{Q}+\overline{P}.Q$ | 7 |
| $= P.(Q+\overline{Q})+\overline{P}.Q$ | 5 |
| $= P.1+\overline{P}.Q$ | 10 |
| $= P+\overline{P}.Q$ | 12 |

*Problem 9* Simplify $F.G.\overline{H}+F.G.H+\overline{F}.G.H$

With reference to *Table 2*:  
$F.G.\bar{H}+F.G.H+\bar{F}.G.H = F.G.(\bar{H}+H)+\bar{F}.G.H$     *Reference* 5  
$\phantom{F.G.\bar{H}+F.G.H+\bar{F}.G.H} = F.G.1+\bar{F}.G.H$     10  
$\phantom{F.G.\bar{H}+F.G.H+\bar{F}.G.H} = F.G+\bar{F}.G.H$     12  
$\phantom{F.G.\bar{H}+F.G.H+\bar{F}.G.H} = G.(F+\bar{F}.H)$     5  

*Problem 10* Simplify $\bar{F}.\bar{G}.H+\bar{F}.G.H+F.\bar{G}.H+F.G.H$

With reference to *Table 2*:     *Reference*  
$\bar{F}.\bar{G}.H+\bar{F}.G.H+F.\bar{G}.H+F.G.H = \bar{G}.H.(\bar{F}+F)+G.H.(\bar{F}+F)$     5  
$\phantom{xxxxxxxxxxxxxxxxxxxxxxxxxxxxxx} = \bar{G}.H.1+G.H.1$     10  
$\phantom{xxxxxxxxxxxxxxxxxxxxxxxxxxxxxx} = \bar{G}.H+G.H$     12  
$\phantom{xxxxxxxxxxxxxxxxxxxxxxxxxxxxxx} = H.(\bar{G}+G)$     5  
$\phantom{xxxxxxxxxxxxxxxxxxxxxxxxxxxxxx} = H.1 = H$     10 and 12  

## C. FURTHER PROBLEMS ON BOOLEAN ALGEBRA AND SWITCHING CIRCUITS

In *Problems 1 to 4*, determine the Boolean expressions and construct truth tables for the switching circuits given.

1   The circuit shown in *Fig 9(a)*.

$[C.(A.B+\bar{A}.B);$ see *Table 4, col. 4*]

2   The circuit shown in *Fig 9(b)*.

$[C.(A.\bar{B}+\bar{A});$ see *Table 4, col. 5*]

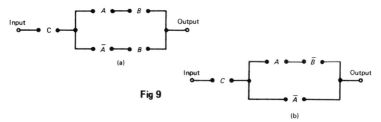

**Fig 9**

TABLE 4

| 1 | 2 | 3 | 4 | 5 | 6 | 7 |
|---|---|---|---|---|---|---|
| $A$ | $B$ | $C$ | $C.(A.B+\bar{A}.B)$ | $C.(A.\bar{B}+\bar{A})$ | $A.B(B.\bar{C}+\bar{B}.C+\bar{A}.B)$ | $C.[b.C.\bar{A}+A.(B+\bar{C})]$ |
| 0 | 0 | 0 | 0 | 0 | 0 | 0 |
| 0 | 0 | 1 | 0 | 1 | 0 | 0 |
| 0 | 1 | 0 | 0 | 0 | 0 | 0 |
| 0 | 1 | 1 | 1 | 1 | 0 | 1 |
| 1 | 0 | 0 | 0 | 0 | 0 | 0 |
| 1 | 0 | 1 | 0 | 1 | 0 | 0 |
| 1 | 1 | 0 | 0 | 0 | 1 | 0 |
| 1 | 1 | 1 | 1 | 0 | 0 | 1 |

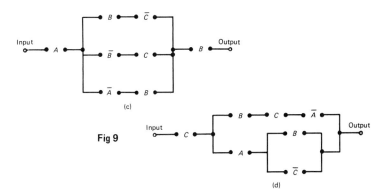

**Fig 9**

3   The circuit shown in *Fig 9(c)*.

$[A.B.(B.\overline{C}+\overline{B}.C+\overline{A}.B)$; see *Table 4, col. 6*]

4   The circuit shown in *Fig 9(d)*.

$[C.[B.C.\overline{A}+A.(B+\overline{C})]$, see *Table 4, col. 7*]

In *Problems 5 to 7*, construct switching circuits to meet the requirements of the Boolean expressions given.

5   $A.C+A.\overline{B}.C+A.B$

[See *Fig 10(a)*]

6   $A.B.C.(A+B+C)$

[See *Fig 10(b)*]

7   $A.(A.\overline{B}.C+B.(A+\overline{C}))$

[See *Fig 10(c)*]

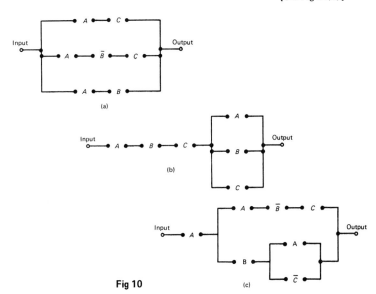

**Fig 10**

In *Problems 8 to 10*, derive the Boolean expressions and construct the switching circuits for the truth table stated.

8  *Table 5, column 4.*

$[\overline{A}.\overline{B}.C+A.B.\overline{C};$ see *Fig 11(a)*]

9  *Table 5, column 5.*

$[\overline{A}.\overline{B}.\overline{C}+\overline{A}.B.C+A.\overline{B}.\overline{C};$ see *Fig 11(b)*]

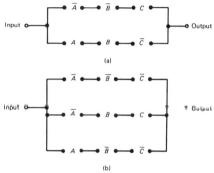

TABLE 5

| 1 | 2 | 3 | 4 | 5 | 6 |
|---|---|---|---|---|---|
| A | B | C |   |   |   |
| 0 | 0 | 0 | 0 | 1 | 1 |
| 0 | 0 | 1 | 1 | 0 | 0 |
| 0 | 1 | 0 | 0 | 0 | 1 |
| 0 | 1 | 1 | 0 | 1 | 0 |
| 1 | 0 | 0 | 0 | 1 | 1 |
| 1 | 0 | 1 | 0 | 0 | 1 |
| 1 | 1 | 0 | 1 | 0 | 0 |
| 1 | 1 | 1 | 0 | 0 | 0 |

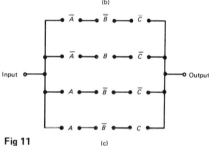

**Fig 11**

10  *Table 5, column 6.*

$[\overline{A}.\overline{B}.\overline{C}+\overline{A}.B.\overline{C}+A.\overline{B}.\overline{C}+A.\overline{B}.C;$ see *Fig 11(c)*]

In *Problems 11 to 13*, show that the relationships given are true by means of a truth table.

11  $(\overline{A}+B).(\overline{A}+C) = \overline{A}+(B.C)$
12  $\overline{A}+A.\overline{B} = \overline{A}+\overline{B}$
13  $A+A.B = A$

In *Problems 14 to 25*, use the laws and rules of Boolean algebra to simplify the expressions given.

14  $\overline{P}.\overline{Q}+\overline{P}.Q$      $[\overline{P}]$
15  $\overline{P}.Q+P.Q$      $[Q]$
16  $\overline{P}.Q+P.Q+\overline{P}.\overline{Q}$      $[\overline{P}+P.Q]$
17  $P.\overline{P}.Q+P.Q.\overline{Q}$      $[0]$
18  $(P+P.Q).(Q+Q.P)$      $[P.Q]$
19  $\overline{F}.\overline{G}.\overline{H}+\overline{F}.\overline{G}.H$      $[\overline{F}.\overline{G}]$
20  $\overline{F}.G.H+F.G.H$      $[G.H]$

21 $\bar{F}.\bar{G}.H+\bar{F}.G.H+F.\bar{G}.H$ $\qquad$ $[H.(\bar{F}+F.\bar{G})]$
22 $F.\bar{G}.\bar{H}+F.G.H+\bar{F}.G.H$ $\qquad$ $[F.\bar{G}.\bar{H}+G.H]$
23 $\bar{F}.\bar{G}.\bar{H}+\bar{F}.\bar{G}.H+F.\bar{G}.\bar{H}+F.\bar{G}.H$ $\qquad$ $[\bar{G}]$
24 $\bar{F}.G.H+\bar{F}.G.H+F.G.H+F.\bar{G}.H$ $\qquad$ $[H.(\bar{F}.G+F)]$
25 $F.\bar{G}.H+F.G.H+F.G.\bar{H}+\bar{F}.G.\bar{H}$ $\qquad$ $[F.H+G.\bar{H}]$

# 20 Introduction to differentiation

## A. MAIN POINTS CONCERNING THE INTRODUCTION TO DIFFERENTIATION

1  **Calculus** is a branch of mathematics involving or leading to calculations dealing with continuously varying functions. Calculus is a subject which falls into two parts: (i) **differential calculus** (or **differentiation**) and (ii) **integral calculus** (or **integration**).

2  In an equation such as $y = 3x^2 + 2x - 5$, $y$ is said to be a function of $x$ and may be written as $y = f(x)$. An equation written in the form $f(x) = 3x^2 + 2x - 5$ is termed **functional notation**. The value of $f(x)$ when $x = 0$ is denoted by $f(0)$, and the value of $f(x)$ when $x = 2$ is denoted by $f(2)$, and so on.
Thus when $f(x) = 3x^2 + 2x - 5$, then $f(0) = 3(0)^2 + 2(0) - 5 = -5$,
and $f(2) = 3(2)^2 + 2(2) - 5 = 11$,
and so on. (See *Problems 1 and 2*.)

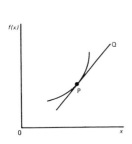

**Fig 1**

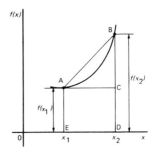

**Fig 2**

3  If a tangent is drawn at a point P on a curve, then the gradient of this tangent is said to be the **gradient of the curve** at P. In *Fig 1*, the gradient of the curve at P is equal to the gradient of the tangent PQ.

4  For the curve shown in *Fig 2*, let the points A and B have co-ordinates $(x_1, y_1)$ and $(x_2, y_2)$ respectively. In functional notation $y_1 = f(x_1)$ and $y_2 = f(x_2)$ as shown.
The gradient of the chord $AB = \dfrac{BC}{AC} = \dfrac{BD - CD}{ED} = \dfrac{f(x_2) - f(x_1)}{(x_2 - x_1)}$

5  For the curve $f(x) = x^2$ shown in *Fig 3*:

  (i) the gradient of chord $AB = \dfrac{f(3) - f(1)}{3 - 1} = \dfrac{9 - 1}{2} = \mathbf{4}$,

  (ii) the gradient of chord $AC = \dfrac{f(2) - f(1)}{2 - 1} = \dfrac{4 - 1}{1} = \mathbf{3}$,

204

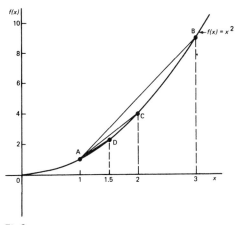

**Fig 3**

(iii) the gradient of chord AD $= \dfrac{f(1.5)-f(1)}{1.5-1} = \dfrac{2.25-1}{0.5} = 2.5$,

(iv) if E is the point on the curve $(1.1, f(1.1))$ then

the gradient of chord AE $= \dfrac{f(1.1)-f(1)}{1.1-1} = \dfrac{1.21-1}{0.1} = 2.1$,

(v) if F is the point on the curve $(1.01, f(1.01))$ then

the gradient of chord AF $= \dfrac{f(1.01)-f(1)}{1.01-1} = \dfrac{1.0201-1}{0.01} = 2.01$.

Thus as point B moves closer and closer to point A the gradient of the chord approaches nearer and nearer to the value 2. This is called the **limiting value** of the gradient of the chord AB and when B coincides with A the chord becomes the tangent to the curve.

6 **Differentiation from first principles**

(i) In *Fig 4*, A and B are two points very close together on a curve, $\delta x$ (delta $x$) and $\delta y$ (delta $y$) representing small increments in the $x$ and $y$ directions respectively.

Gradient of chord AB $= \dfrac{\delta y}{\delta x}$

However
$\delta y = f(x+\delta x) - f(x)$
Hence
$\dfrac{\delta y}{\delta x} = \dfrac{f(x+\delta x)-f(x)}{\delta x}$

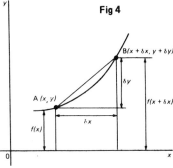

**Fig 4**

As $\delta x$ approaches zero, $\delta y/\delta x$ approaches a limiting value and the gradient of the chord approaches the gradient of the tangent at A.

(ii) When determining the gradient of a tangent to a curve there are two notations used. The gradient of the curve at A in *Fig 4* can either be written as

$$\lim_{\delta x \to 0} \frac{\delta y}{\delta x} \quad \text{or} \quad \lim_{\delta x \to 0} \left\{ \frac{f(x+\delta x)-f(x)}{\delta x} \right\}$$

In **Leibniz notation**, $\quad \dfrac{dy}{dx} = \lim\limits_{\delta x \to 0} \dfrac{\delta y}{\delta x}$

In **functional notation**, $f'(x) = \lim\limits_{\delta x \to 0} \left\{ \dfrac{f(x+\delta x)-f(x)}{\delta x} \right\}$

(iii) $dy/dx$ is the same as $f'(x)$ and is called the **differential coefficient** or the **derivative**. The process of finding the differential coefficient is called **differentiation**. (See *Problems 3 to 7*.)

Summarising, the differential coefficient, $\dfrac{dy}{dx} = f'(x) = \lim\limits_{\delta x \to 0} \dfrac{\delta y}{\delta x}$

$$= \lim\limits_{\delta x \to 0} \left\{ \frac{f(x+\delta x)-f(x)}{\delta x} \right\}$$

7 From differentiation by first principles, a general rule for differentiating $ax^n$ emerges where $a$ and $n$ are any constants.

This rule is: if $y = ax^n$ then $\dfrac{dy}{dx} = anx^{n-1}$

or, if $f(x) = ax^n$ then $f'(x) = anx^{n-1}$

When differentiating, results can be expressed in a number of ways. For example:

(i) if $y = 3x^2$ then $\dfrac{dy}{dx} = 6x$,

(ii) if $f(x) = 3x^2$ then $f'(x) = 6x$,

(iii) the differential coefficient of $3x^2$ is $6x$,

(iv) the derivative of $3x^2$ is $6x$, and

(v) $\dfrac{d}{dx}(3x^2) = 6x$

8 *Fig 5(a)* shows a graph of $y = \sin \theta$. The gradient is continually changing as the curve moves from 0 to A to B to C to D. The gradient, given by $dy/d\theta$, may be plotted in a corresponding position below $y = \sin \theta$, as shown in *Fig 5(b)*.
  (i) At 0, the gradient is positive and is at its steepest. Hence 0′ is a maximum positive value.
  (ii) Between 0 and A the gradient is positive but is decreasing in value until at A the gradient is zero, shown as A′.
  (iii) Between A and B the gradient is negative but is increasing in value until at B the gradient is at its steepest. Hence B′ is a maximum negative value.
  (iv) If the gradient of $y = \sin \theta$ is further investigated between B and C and C and D then the resulting graph of $dy/d\theta$ is seen to be a cosine wave. Hence the rate of change of $\sin \theta$ is $\cos \theta$,
  i.e., **if $y = \sin \theta$ then $dy/d\theta = \cos \theta$**.

9 **Maximum and minimum points**

In *Fig 6*, the gradient (or rate of change) of the curve changes from positive

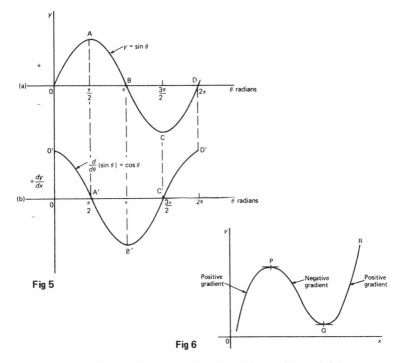

**Fig 5**

**Fig 6**

between 0 and P to negative between P and Q and then positive again between Q and R. At point P the gradient is zero and, as $x$ increases, the gradient of the curve changes from positive just before P to negative just after. Such a point is called a **maximum point** and appears as the 'crest of a wave'. At point Q, the gradient is also zero and, as $x$ increases, the gradient of the curve changes from negative just before Q to positive just after. Such a point is called a **minimum value**, and appears as the 'bottom of a valley'. Points such as P and Q are given the general name of **turning points**.

## B. WORKED PROBLEMS ON THE INTRODUCTION TO DIFFERENTIATION

*Problem 1* If $f(x) = 4x^2 - 3x + 2$ find $f(0), f(3), f(-1)$ and $f(3) - f(-1)$.

$f(x) = 4x^2 - 3x + 2$
$f(0) = 4(0)^2 - 3(0) + 2 = \mathbf{2}$
$f(3) = 4(3)^2 - 3(3) + 2 = 36 - 9 + 2 = \mathbf{29}$
$f(-1) = 4(-1)^2 - 3(-1) + 2 = 4 + 3 + 2 = \mathbf{9}$
$f(3) - f(-1) = 29 - 9 = \mathbf{20}.$

*Problem 2* Given that $f(x) = 5x^2+x-7$ determine (i) $f(2) \div f(1)$, (ii) $f(3+a)$, (iii) $f(3+a)-f(3)$ and (iv) $\dfrac{f(3+a)-f(3)}{a}$

$f(x) = 5x^2+x-7$.

(i) $f(2) = 5(2)^2+2-7 = 15$
 $f(1) = 5(1)+1-7 = -1$
 $f(2) \div f(1) = \dfrac{15}{-1} = -15$

(ii) $f(3+a) = 5(3+a)^2+(3+a)-7$
 $= 5(9+6a+a^2)+(3+a)-7$
 $= 45+30a+5a^2+3+a-7$
 $= 41+31a+5a^2$

(iii) $f(3) = 5(3)^2+3-7 = 41$
 $f(3+a)-f(3) = (41+31a+5a^2)-(41) = 31a+5a^2$

(iv) $\dfrac{f(3+a)-f(3)}{a} = \dfrac{31a+5a^2}{a} = 31+5a$

*Problem 3* Differentiate from first principles $f(x) = x^2$ and determine the value of the gradient of the curve at $x = 2$.

To 'differentiate from first principles' means 'to find $f'(x)$ by using the expression
$f'(x) = \underset{\delta x \to 0}{\text{limit}} \left\{ \dfrac{f(x+\delta x)-f(x)}{\delta x} \right\}$, (see para. 6)

$f(x) = x^2$
Substituting $(x+\delta x)$ for $x$ gives $f(x+\delta x) = (x+\delta x)^2 = x^2+2x\delta x+\delta x^2$

Hence $f'(x) = \underset{\delta x \to 0}{\text{limit}} \left\{ \dfrac{(x^2+2x\delta x+\delta x^2)-(x^2)}{\delta x} \right\}$

$= \underset{\delta x \to 0}{\text{limit}} \left\{ \dfrac{2x\delta x+\delta x^2}{\delta x} \right\}$

$= \underset{\delta x \to 0}{\text{limit}} \{ 2x+\delta x \}$

As $\delta x \to 0, \{ 2x+\delta x \} \to \{ 2x+0 \}$
Thus $f'(x) = 2x$, i.e. the differential coefficient of $x^2$ is $2x$.
At $x = 2$, the gradient of the curve, $f'(x) = 2(2) = 4$.

*Problem 4* Find the differential coefficient of $y = 5x$

By definition, $\dfrac{dy}{dx} = f'(x) = \underset{\delta x \to 0}{\text{limit}} \left\{ \dfrac{f(x+\delta x)-f(x)}{\delta x} \right\}$

The function being differentiated is $y = f(x) = 5x$.
Substituting $(x+\delta x)$ for $x$ gives $f(x+\delta x) = 5(x+\delta x) = 5x + 5\delta x$

Hence $\dfrac{dy}{dx} = f'(x) = \underset{\delta x \to 0}{\text{limit}} \left\{ \dfrac{(5x+5\delta x)-(5x)}{\delta x} \right\}$

$= \underset{\delta x \to 0}{\text{limit}} \left\{ \dfrac{5\delta x}{\delta x} \right\} = \underset{\delta x \to 0}{\text{limit}} \{5\}$

Since the term $\delta x$ does not appear in $\{5\}$ the limiting value as $\delta x \to 0$ of $\{5\}$ is 5.

Thus $\dfrac{dy}{dx} = 5$, i.e. the differential coefficient of $5x$ is 5.

The equation $y = 5x$ represents a straight line of gradient 5 (see chapter 13).

The 'differential coefficient' (i.e. $\dfrac{dy}{dx}$ or $f'(x)$) means 'the gradient of the curve', and since the slope of the line $y = 5x$ is 5 this result can be obtained by inspection. Hence, in general, if $y = kx$, (where $k$ is a constant), then the slope of the line is $k$ and $\dfrac{dy}{dx}$ or $f'(x) = k$.

*Problem 5* Find the derivative of $y = 8$.

$y = f(x) = 8$.
Since there are no $x$-values in the original equation, substituting $(x+\delta x)$ for $x$ still gives $f(x+\delta x) = 8$.

Hence $\dfrac{dy}{dx} = f'(x) = \underset{\delta x \to 0}{\text{limit}} \left\{ \dfrac{f(x+\delta x)-f(x)}{\delta x} \right\} = \underset{\delta x \to 0}{\text{limit}} \left\{ \dfrac{8-8}{\delta x} \right\} = 0$

Thus, when $y = 8$, $\dfrac{dy}{dx} = 0$.

The equation $y = 8$ represents a straight horizontal line and the gradient of a horizontal line is zero, hence the result could have been determined by inspection. 'Finding the derivative' means 'finding the gradient' hence, in general, for any horizontal line if $y = k$ (where $k$ is a constant) then $dy/dx = 0$.

*Problem 6* Differentiate from first principles $f(x) = 2x^3$.

Substituting $(x+\delta x)$ for $x$ gives $f(x+\delta x) = 2(x+\delta x)^3$
$= 2(x+\delta x)(x^2+2x\delta x+\delta x^2)$
$= 2(x^3+3x^2\delta x+3x\delta x^2+\delta x^3)$
$= 2x^3+6x^2\delta x+6x\delta x^2+2\delta x^3)$

$f'(x) = \underset{\delta x \to 0}{\text{limit}} \left\{ \dfrac{f(x+\delta x)-f(x)}{\delta x} \right\}$

$= \underset{\delta x \to 0}{\text{limit}} \left\{ \dfrac{(2x^3+6x^2\delta x+6x\delta x^2+2\delta x^3)-(2x^3)}{\delta x} \right\}$

$$= \underset{\delta x \to 0}{\text{limit}} \left\{ \frac{6x^2 \delta x + 6x \delta x^2 + 2 \delta x^3}{\delta x} \right\}$$

$$= \underset{\delta x \to 0}{\text{limit}} \{ 6x^2 + 6x \delta x + 2 \delta x^2 \}$$

Hence $f'(x) = 6x^2$, i.e. the differential coefficient of $2x^3$ is $6x^2$

*Problem 7* Find the differential coefficient of $y = 4x^2 + 5x - 3$ and determine the gradient of the curve at $x = -3$.

$$\begin{aligned}
y = f(x) &= 4x^2 + 5x - 3 \\
f(x+\delta x) &= 4(x+\delta x)^2 + 5(x+\delta x) - 3 \\
&= 4(x^2 + 2x\delta x + \delta x^2) + 5x + 5\delta x - 3 \\
&= 4x^2 + 8x\delta x + 4\delta x^2 + 5x + 5\delta x - 3
\end{aligned}$$

$$\begin{aligned}
\frac{dy}{dx} = f'(x) &= \underset{\delta x \to 0}{\text{limit}} \left\{ \frac{f(x+\delta x) - f(x)}{\delta x} \right\} \\
&= \underset{\delta x \to 0}{\text{limit}} \left\{ \frac{(4x^2 + 8x\delta x + 4\delta x^2 + 5x + 5\delta x - 3) - (4x^2 + 5x - 3)}{\delta x} \right\} \\
&= \underset{\delta x \to 0}{\text{limit}} \left\{ \frac{8x\delta x + 4\delta x^2 + 5\delta x}{\delta x} \right\} \\
&= \underset{\delta x \to 0}{\text{limit}} \{ 8x + 4\delta x + 5 \}
\end{aligned}$$

i.e. $\dfrac{dy}{dx} = f'(x) = 8x + 5$

At $x = -3$, the gradient of the curve $= \dfrac{dy}{dx} = f'(x) = 8(-3) + 5 = -19$

(Each of the results obtained in *Problems 3 to 7* may be deduced by using the general rule stated in para. 7).

## C. FURTHER PROBLEMS ON THE INTRODUCTION TO DIFFERENTIATION

1. If $f(x) = 6x^2 - 2x + 1$ find $f(0), f(1), f(2), f(-1)$ and $f(-3)$.

    [1, 5, 21, 9, 61]

2. If $f(x) = 2x^2 + 5x - 7$ find $f(1), f(2), f(-1)$ and $f(2) - f(-1)$.

    [0, 11, −10, 21]

3. Given $f(x) = 3x^3 + 2x^2 - 3x + 2$ prove that $f(1) = \dfrac{1}{7} f(2)$.

4. If $f(x) = -x^2 + 3x + 6$ find $f(2), f(2+a), f(2+a) - f(2)$ and $\dfrac{f(2+a) - f(2)}{a}$

    [8; $-a^2 - a + 8$; $-a^2 - a$; $-a - 1$]

5. Plot the curve $f(x) = 4x^2 - 1$ for values of $x$ from $x = -1$ to $x = +4$. Label the co-ordinates $(3, f(3))$ and $(1, f(1))$ as J and K respectively. Join points J and K to form the chord JK. Determine the gradient of chord JK. By moving K nearer and nearer to J determine the gradient of the tangent of the curve at J.

    [16; 8]

In *Problems 6 to 17*, differentiate from first principles.

6. $y = x$ [1]

7. $y = 7x$ [7]
8. $y = 4x^2$ [$8x$]
9. $y = 5x^3$ [$15x^2$]
10. $y = -2x^2+3x-12$ [$-4x+3$]
11. $y = 23$ [0]
12. $f(x) = 9x$ [9]
13. $f(x) = \dfrac{2x}{3}$ [$\dfrac{2}{3}$]
14. $f(x) = 9x^2$ [$18x$]
15. $f(x) = -7x^3$ [$-21x^2$]
16. $f(x) = x^2+15x-4$ [$2x+15$]
17. $f(x) = 4$ [0]
18. Determine $\dfrac{d}{dx}(4x^3)$ from first principles. [$12x^2$]
19. Find $\dfrac{d}{dx}(3x^2+5)$ from first principles. [$6x$]
20. Using the general rule for $ax^n$ (see para. 7) check the results of *Problems 6 to 19*.
21. Differentiate from first principles $f(x) = 6x^2-3x+5$ and find the gradient of the curve at (a) $x = -1$ and (b) $x = 2$.

    [$12x-3$; (a) $-15$; (b) $21$]
22. Find the differential coefficient of $y = 2x^3+3x^2-4x-1$ and determine the gradient of the curve at $x = 2$.

    [$6x^2+6x-4$; $32$]
23. Determine the derivative of $y = -2x^3+4x+7$ and determine the gradient of the curve at $x = -1.5$.

    [$-6x^2+4$; $-9.5$]
24. Show graphically that the rate of change of $\sin\theta$ is $\cos\theta$.
25. Define (a) a maximum point and (b) a minimum point.

# 21 Introduction to integration

## A. MAIN POINTS CONCERNING AN INTRODUCTION TO INTEGRATION

1. The process of integration reverses the process of differentiation. In differentiation, if $f(x) = 2x^2$ then $f'(x) = 4x$. Thus the integral of $4x$ is $2x^2$, i.e. integration is the process of moving from $f'(x)$ to $f(x)$. By similar reasoning, the integral of $2t$ is $t^2$.

2. Integration is a process of summation or adding parts together and an elongated S, shown as $\int$, is used to replace the words 'the integral of'. Hence, from para 1, $\int 4x = 2x^2$ and $\int 2t$ is $t^2$.

3. In differentiation, the differential coefficient $dy/dx$ indicates that a function of $x$ is being differentiated with respect to $x$, the $dx$ indicating that it is 'with respect to $x$'. In integration the variable of integration is shown by adding $d$ (the variable) after the function to be integrated.

   Thus $\int 4x \, dx$ means 'the integral of $4x$ with respect to $x$',
   and $\int 2t \, dt$ means 'the integral of $2t$ with respect to $t$'.

4. As stated in para. 1, the differential coefficient of $2x^2$ is $4x$, hence $\int 4x \, dx = 2x^2$. However, the differential coefficient of $2x^2 + 7$ is also $4x$. Hence $\int 4x \, dx$ is also equal to $2x^2 + 7$. To allow for the possible presence of a constant, whenever the process of integration is performed, a constant '$c$' is added to the result.
   Thus $\int 4x \, dx = 2x^2 + c$ and $\int 2t \, dt = t^2 + c$
   '$c$' is called the **arbitrary constant of integration**.

5. The general solution of integrals of the form $\int ax^n dx$, where $a$ and $n$ are constants and $n \neq -1$ is given by:

   $$\int ax^n dx = \frac{ax^{n+1}}{n+1} + c$$

   Using this rule gives: $\int 3x^4 dx = \frac{3x^{4+1}}{4+1} + c = \frac{3}{5}x^5 + c$

   and $\int \frac{4}{9} t^3 \, dt = \frac{4}{9}\left(\frac{t^{3+1}}{3+1}\right) + c = \frac{4}{9}\left(\frac{t^4}{4}\right) + c = \frac{1}{9}t^4 + c$

   Both of these results may be checked by differentiation.

6. Integrals containing an arbitrary constant $c$ in their results are called **indefinite integrals** since their precise value cannot be determined without further information. **Definite integrals** are those in which limits are applied. If an expression is written as $[x]_a^b$, $b$ is called the upper limit and $a$ the lower limit. The operation of applying the limits is defined as $[x]_a^b = (b) - (a)$. The increase in the value of the integral $x^2$ as $x$ increases from 1 to 3 is written as $\int_1^3 x^2 dx$.

   Applying the limits gives: $\int_1^3 x^2 dx = \left[\frac{x^3}{3} + c\right]_1^3$
   $= \left\{\frac{(3)^3}{3} + c\right\} - \left\{\frac{(1)^3}{3} + c\right\} = (9+c) - \left(\frac{1}{3} + c\right) = 8\frac{2}{3}$

Note that the 'c' term always cancels out when limits are applied and it need not be shown with definite integrals.

7   The area shown shaded in *Fig 1* may be determined using approximate methods such as the trapezoidal rule, the mid-ordinate rule or Simpson's rule (see chapter 6), or precisely by using integration.

The shaded area in *Fig 1* is given by:

$$\int_a^b y \, dx = \int_a^b f(x) dx.$$

**Fig 1**

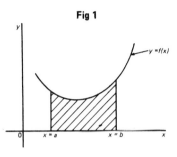

There are several instances in engineering and science where the area beneath a curve needs to be accurately determined. For example, the area between limits of a:

velocity/time graph gives distance travelled, force/distance graph gives work done, voltage/current graph gives power, and so on.

8   Should a curve drop below the $x$-axis, then $y$ $(= f(x))$ becomes negative and $\int f(x) dx$ is negative. When determining such areas by integration, a negative sign is placed before the integral. For the curve shown in *Fig 2*, the total shaded area is given by (area E + area F + area G).

By integration, **total shaded area** = $\int_a^b f(x) dx - \int_b^c f(x) dx + \int_c^d f(x) dx.$

(Note that this is **not** the same as $\int_a^d f(x) \, dx$.)

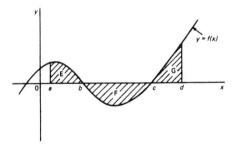

**Fig 2**

It is usually necessary to sketch a curve in order to check whether it crosses the $x$-axis.

9   From chapter 20, the differential coefficient of $\sin \theta$ is $\cos \theta$. Since integration is the reverse process of differentiation then:

$$\int \cos \theta \, d\theta = \sin \theta + c$$

## B. WORKED PROBLEMS ON THE INTRODUCTION TO INTEGRATION

**Problem 1** Determine (a) $\int 3x^2\, dx$; (b) $\int 2t^3\, dt$

The general rule is $\int ax^n\, dx = \dfrac{ax^{n+1}}{n+1} + c$

(a) When $a = 3$ and $n = 2$ then $\int 3x^2\, dx = \dfrac{3x^{2+1}}{2+1} + c = \dfrac{3x^3}{3} + c = x^3 + c.$

(b) When $a = 2$ and $n = 3$ then $\int 2t^3\, dt = \dfrac{2t^{3+1}}{3+1} + c = \dfrac{2}{4}t^4 + c = \dfrac{1}{2}t^4 + c.$

Each of these results may be checked by differentiating them.

**Problem 2** Determine (a) $\int 8\, dx$; (b) $\int \dfrac{2}{3}x\, dx$

(a) $\int 8\, dx$ is the same as $\int 8x^0\, dx$, and, using the general rule

when $a = 8$ and $n = 0$ gives $\int 8x^0\, dx = \dfrac{8x^{0+1}}{0+1} + c = 8x + c.$

In general, if $k$ is a constant then $\int k\, dx = kx + c.$

(b) When $a = \dfrac{2}{3}$ and $n = 1$ then $\int \dfrac{2}{3}x\, dx = \left(\dfrac{2}{3}\right) \dfrac{x^{1+1}}{(1+1)} + c = \left(\dfrac{2}{3}\right) \dfrac{x^2}{2} + c$

$$= \dfrac{1}{3}x^2 + c$$

**Problem 3** Determine $\int (2 + \dfrac{5}{7}x - 6x^2)\, dx$

$\int \left(2 + \dfrac{5}{7}x - 6x^2\right) dx$ may be written as $\int 2\, dx + \int \dfrac{5}{7}x\, dx - \int 6x^2\, dx$
i.e., each term is integrated separately. (This splitting up of terms only applies for addition and subtraction.)

Hence $\int \left(2 + \dfrac{5}{7}x - 6x^2\right) dx = \dfrac{2x^{0+1}}{0+1} + \left(\dfrac{5}{7}\right) \dfrac{x^{1+1}}{(1+1)} - \dfrac{6x^{2+1}}{(2+1)} + c.$

$$= 2x + \left(\dfrac{5}{7}\right) \dfrac{x^2}{2} - 6\dfrac{x^3}{3} + c = 2x + \dfrac{5}{14}x^2 - 2x^3 + c.$$

Note that when an integral contains more than one term there is no need to have an arbitrary constant for each; just a single constant at the end is sufficient.

**Problem 4** Determine (a) $\int \sqrt{x}\, dx$; (b) $\int \dfrac{3}{x^2}\, dx$

When $n$ is fractional or negative the general rule for integrals of the form $\int ax^n\, dx$ can still be applied.

(a) $\int \sqrt{x}\, dx = \int x^{\frac{1}{2}} dx$

Using the general rule, where $a = 1$ and $n = \frac{1}{2}$ gives:

$\int x^{\frac{1}{2}} dx = \frac{(1)x^{\frac{1}{2}+1}}{\frac{1}{2}+1} + c = \frac{x^{\frac{3}{2}}}{\frac{3}{2}} + c = \frac{2}{3}x^{\frac{3}{2}} + c = \frac{2}{3}\sqrt{x^3} + c$

(b) $\int \frac{3}{x^2} dx = \int 3x^{-2} dx$

Using the general rule, where $a = 3$ and $n = -2$ gives:

$\int 3x^{-2} dx = \frac{3x^{-2+1}}{-2+1} + c = \frac{3x^{-1}}{-1} + c = -3x^{-1} + c = -\frac{3}{x} + c.$

***Problem 5*** Determine (a) $\int \left(\frac{x^3 - 2x}{3x}\right) dx$; (b) $\int (1-x)^2 dx$

(a) $\int \left(\frac{x^3 - 2x}{3x}\right) dx = \int \left(\frac{x^3}{3x} - \frac{2x}{3x}\right) dx = \int \left(\frac{x^2}{3} - \frac{2}{3}\right) dx$

$= \left(\frac{1}{3}\right) \frac{x^{2+1}}{(2+1)} - \frac{2}{3}x + c = \frac{1}{9}x^3 - \frac{2}{3}x + c$

(b) $\int (1-x)^2 dx = \int (1 - 2x + x^2) dx = \frac{(1)x^{0+1}}{0+1} - \frac{(2)x^{1+1}}{(1+1)} + \frac{x^{2+1}}{2+1} + c$

$= x - x^2 + \frac{1}{3}x^3 + c$

This problem shows that functions often have to be rearranged into the standard form of $\int ax^n dx$ before it is possible to integrate them.

***Problem 6*** Evaluate (a) $\int_1^2 4x\, dx$; (b) $\int_{-2}^3 (5 - x^2) dx$

(a) $\int_1^2 4x\, dx = \left[\frac{4x^2}{2}\right]_1^2 = [2x^2]_1^2 = \{2(2)^2\} - \{2(1)^2\} = 8 - 2 = 6$

(b) $\int_{-2}^3 (5 - x^2) dx = \left[5x - \frac{x^3}{3}\right]_{-2}^3 = \left\{5(3) - \frac{(3)^3}{3}\right\} - \left\{5(-2) - \frac{(-2)^3}{3}\right\}$

$= \{15 - 9\} - \left\{-10 - \frac{-8}{3}\right\} = 6 + 10 - \frac{8}{3} = 13\frac{1}{3}$

***Problem 7*** Evaluate (a) $\int_0^2 x(3 + 2x) dx$; (b) $\int_{-1}^1 \frac{(x^4 - 5x^2 + x)}{x} dx$

(a) $\int_0^2 x(3 + 2x) dx = \int_0^2 (3x + 2x^2) dx = \left[\frac{3x^2}{2} + \frac{2x^3}{3}\right]_0^2$

$= \left\{\frac{3(2)^2}{2} + \frac{2(2)^3}{3}\right\} - \{0 + 0\} = 6 + \frac{16}{3} = 11\frac{1}{3}$

(b) $\int_{-1}^{1} \frac{(x^4 - 5x^2 + x)}{x} dx = \int_{-1}^{1} (x^3 - 5x + 1) dx = \left[\frac{x^4}{4} - \frac{5x^2}{2} + x\right]_{-1}^{1}$

$= \left\{\frac{1}{4} - \frac{5}{2} + 1\right\} - \left\{\frac{(-1)^4}{4} - \frac{5(-1)^2}{2} + (-1)\right\}$

$= \left(\frac{1}{4} - \frac{5}{2} + 1\right) - \left(\frac{1}{4} - \frac{5}{2} - 1\right) = 2$

*Problem 8* Determine the area enclosed by $y = 2x+3$, the x-axis and ordinates $x = 0$ and $x = 3$.

$y = 2x+3$ is a straight line graph as shown in *Fig 3*, where the area enclosed by $y = 2x+3$, the x-axis and ordinates $x = 0$ and $x = 3$ is shown shaded.

By integration, shaded area $= \int_{0}^{3} y\, dx = \int_{0}^{3} (2x+3) dx$

$= \left[\frac{2x^2}{2} + 3x\right]_{0}^{3} = (9+9)-(0+0)$

$= 18$ square units

(This answer may be checked since the shaded area is a trapezium.

Area of trapezium $= \frac{1}{2}$(sum of parallel sides)(perpendicular distance between parallel sides)

$= \frac{1}{2}(3+9)(3) = 18$ square units.)

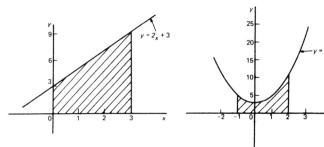

Fig 3        Fig 4

*Problem 9* Sketch the graph of $y = 2x^2+3$ from $x = -2$ to $x = 3$. Find, by integration, the area enclosed by the curve, the x-axis and ordinates $x = -1$ and $x = 2$.

Since $2x^2+3$ is a quadratic expression, the curve $y = 2x^2+3$ is a parabola, cutting the y-axis at $y = 3$, as shown in Fig 4. The area required is shown shaded.

By integration, shaded area $= \int_{-1}^{2} y\, dx = \int_{-1}^{2} (2x^2+3)dx$

$= \left[\frac{2x^3}{3} + 3x\right]_{-1}^{2}$

$= \left\{\frac{2(2)^3}{3} + 3(2)\right\} - \left\{\frac{2(-1)^3}{3} + 3(-1)\right\}$

$= \left(\frac{16}{3}+6\right) - \left(-\frac{2}{3}-3\right) = \left(11\frac{1}{3}\right) - \left(-3\frac{2}{3}\right)$

$= 15$ square units

*Problem 10* Sketch the graph $y = x^3 - 2x^2 - 5x + 6$ between $x = -2$ and $x = 3$, and determine the area enclosed by the curve and the x-axis.

A table of values is produced and the graph sketched, as shown in *Fig 5*, where the area enclosed by the curve and the x-axis is shown shaded.

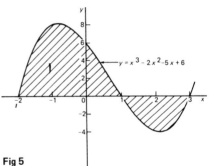

Fig 5

| $x$     | -2 | -1 | 0 | 1  | 2   | 3   |
|---------|----|----|---|----|-----|-----|
| $x^3$   | -8 | -1 | 0 | 1  | 8   | 27  |
| $-2x^2$ | -8 | -2 | 0 | -2 | -8  | -18 |
| $-5x$   | 10 | 5  | 0 | -5 | -10 | -15 |
| $+6$    | 6  | 6  | 6 | 6  | 6   | 6   |
| $y$     | 0  | 8  | 6 | 0  | -4  | 0   |

The shaded area $= \int_{-2}^{1} y\, dx - \int_{1}^{3} y\, dx$, the minus sign before the second integral being necessary since the enclosed area is below the x-axis.

Hence shaded area = $\int_{-2}^{1} (x^3-2x^2-5x+6)dx - \int_{1}^{3} (x^3-2x^2-5x+6)dx$

$\int_{-2}^{1} (x^3-2x^2-5x+6)dx = \left[\frac{x^4}{4} - \frac{2x^3}{3} - \frac{5x^2}{2} + 6x\right]_{-2}^{1}$

$= \left\{\frac{1}{4} - \frac{2}{3} - \frac{5}{2} + 6\right\} - \left\{\frac{(-2)^4}{4} - \frac{2(-2)^3}{3} - \frac{5(-2)^2}{2} + 6(-2)\right\}$

$= \left(3\frac{1}{12}\right) - \left(-12\frac{2}{3}\right) = 15\frac{3}{4}$ square units

$\int_{1}^{3} (x^3-2x^2-5x+6)dx = \left[\frac{x^4}{4} - \frac{2x^3}{3} - \frac{5x^2}{2} + 6x\right]_{1}^{3}$

$= \left\{\frac{81}{4} - 18 - \frac{45}{2} + 18\right\} - \left\{\frac{1}{4} - \frac{2}{3} - \frac{5}{2} + 6\right\}$

$= \left(-2\frac{1}{4}\right) - \left(3\frac{1}{12}\right) = -5\frac{1}{3}$ square units

Hence, shaded area = $\left(15\frac{3}{4}\right) - \left(-5\frac{1}{3}\right) = 21\frac{1}{12}$ square units

*Problem 11* Determine the area enclosed by the curve $y = \cos\theta$, the ordinates $\theta = 0$ and $\theta = \pi/2$ and the $\theta$-axis.

The graph of $y = \cos\theta$ is shown in *Fig 6* and the required area is shown shaded.

The shaded area $= \int_{0}^{\pi/2} \cos\theta \, d\theta$

$= [\sin\theta]_{0}^{\pi/2}$

$= \sin\frac{\pi}{2} - \sin 0 = 1$ square unit

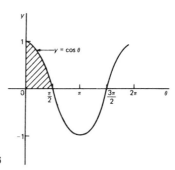

Fig 6

*Problem 12* Determine the area enclosed by the curve $y = 3x^2+6$, the x-axis and ordinates $x = 1$ and $x = 4$, (a) by integration, (b) by the trapezoidal rule, (c) by the mid-ordinate rule, and (d) by Simpson's rule.

The curve $y = 3x^2+6$ is shown plotted in *Fig 7*

(a) **By integration**, shaded area $= \int_{1}^{4} y \, dx = \int_{1}^{4} (3x^2+6)dx$

$= [x^3+6x]_{1}^{4} = \{(4)^3+6(4)\} - \{(1)^3+6(1)\}$

$= (64+24) - (1+6) = 81$ **square units**

| x | 0 | 1.0 | 1.5 | 2.0 | 2.5 | 3.0 | 3.5 | 4.0 |
|---|---|-----|------|-----|------|-----|------|-----|
| $y = 3x^2 + 6$ | 6.0 | 9.0 | 12.75 | 18.0 | 24.75 | 33.0 | 42.75 | 54.0 |

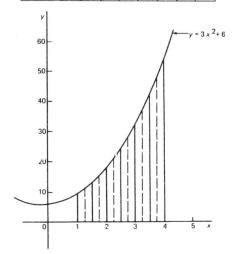

Fig 7

(b) **By the trapezoidal rule** (see chapter 6, para. 1(b)):

Area = (width of interval)$\left\{\dfrac{1}{2}(\text{first + last ordinates}) + \text{sum of remaining ordinates}\right\}$

Selecting 6 intervals, each of width 0.5, gives:

Area = $(0.5)\left\{\dfrac{1}{2}(9+54)+12.75+18+24.75+33+42.75\right\}$

= **81.375 square units**

(c) **By the mid-ordinate rule** (see chapter 6, para. 1(c)):

Area = (width of interval)(sum of mid-ordinates)

Selecting 6 intervals, each of width 0.5, gives the mid-ordinates as shown by the broken line in *Fig 7*.
Thus, Area = $(0.5)(11+15+21+28.5+37.5+48)$
= **80.5 square units**

(d) **By Simpson's rule** (see chapter 6, para 1(d)):

Area = $\dfrac{1}{3}$(width of interval)$\left\{\begin{pmatrix}\text{first + last}\\\text{ordinate}\end{pmatrix} + 4\begin{pmatrix}\text{sum of even}\\\text{ordinates}\end{pmatrix} + 2\begin{pmatrix}\text{sum of remaining}\\\text{odd ordinates}\end{pmatrix}\right\}$

Selecting 6 intervals, each of width 0.5, gives:

Area = $\dfrac{1}{3}(0.5)[(9+54)+4(12.75+24.75+42.75)+2(18+33)]$

= **81 square units**

Integration gives the correct result for the area under a curve. Simpson's rule is seen to be the most accurate of the three approximate methods in this case.

## C. FURTHER PROBLEMS ON THE INTRODUCTION TO INTEGRATION

Determine the indefinite integrals in *Problems 1 to 6*.

1  (a) $\int 5 dx$; (b) $\int 7x\, dx$

$$\left[\text{(a) } 5x+c; \text{(b) } \frac{7x^2}{2} + c\right]$$

2  (a) $\int \frac{2}{3} x^2 dx$; (b) $\int \frac{4}{5} x^3 dx$

$$\left[\text{(a) } \frac{2}{9}x^3 + c; \text{(b) } \frac{1}{5}x^4 + c\right]$$

3  (a) $\int (3+2x-4x^2) dx$; (b) $3\int (x-5x^2) dx$

$$\left[\text{(a) } 3x+x^2 - \frac{4x^3}{3} + c; \text{(b) } 3\left(\frac{x^2}{2} - \frac{5x^3}{3}\right) + c\right]$$

4  (a) $\int \frac{3x^2 - 5x}{2x} dx$; (b) $\int (2+x)^2 dx$

$$\left[\text{(a) } \frac{3}{4}x^2 - \frac{5}{2}x + c; \text{(b) } 4x+2x^2 + \frac{x^3}{3} + c\right]$$

5  (a) $\int \frac{2}{x^3} dx$; (b) $\int \frac{9}{4x^2} dx$

$$\left[\text{(a) } \frac{-1}{x^2} + c; \text{(b) } \frac{-9}{4x} + c\right]$$

(6)  (a) $\int 7\sqrt{x^3}\, dx$; (b) $\int \frac{4}{\sqrt{x}} dx$

$$\left[\text{(a) } \frac{14}{5}\sqrt{x^5} + c; \text{(b) } 8\sqrt{x} + c\right]$$

Evaluate the definite integrals in *Problems 7 to 10*.

7  (a) $\int_1^4 3x^2 dx$; (b) $\int_{-1}^2 2x\, dx$

$$[\text{(a) } 80; \text{(b) } 3]$$

8  (a) $\int_0^2 (5x-2x^2) dx$; (b) $\int_1^3 (x^2-4x+3) dx$

$$\left[\text{(a) } 4\frac{2}{3}; \text{(b) } -1\frac{1}{3}\right]$$

9  (a) $\int_0^\pi 3 \cos \theta\, d\theta$; (b) $\int_0^{\pi/2} 5 \cos \theta\, d\theta$

$$[\text{(a) } 0; \text{(b) } 5]$$

10  (a) $\int_1^2 (3-x)^2 dx$; (b) $\int_{-2}^{-1} \frac{2x+x^2}{x} dx$

$$\left[\text{(a) } 2\frac{1}{3}; \text{(b) } 4\frac{1}{2}\right]$$

11 Show by integration that the area of a rectangle formed by the line $y = 4$, the ordinate $x = 1$ and $x = 6$ and the x-axis is 20 square units.

12 Show by integration that the area of the triangle formed by the line $y = 3x$, the ordinates $x = 0$ and $x = 5$ and the x-axis is 37.5 square units.

13 Sketch the curve $y = x^2 + 5$ between $x = -1$ and $x = 4$. Find by integration the area enclosed by the curve the x-axis and ordinates $x = 1$ and $x = 3$. Use an approximate method to find the area and compare your result with that obtained by integration.

$$\left[18\frac{2}{3} \text{ square units}\right]$$

14 Determine the area enclosed by $y = 2x^3$, the x-axis and ordinates $x = 0$ and $x = 2$;

[8 square units]

15 Determine the area enclosed between the curve $y = 6-x-x^2$ and the x-axis.

$\left[20\frac{5}{6} \text{ square units}\right]$

16 Sketch the curve $y = x(x-1)(x-3)$ and use Simpson's rule to find the area enclosed by the curve and the x-axis. Compare your answer with the true area obtained by integration.

$\left[3\frac{1}{12} \text{ square units}\right]$

17 Find the area enclosed by the curve $y = x^2+x-6$, the x-axis and ordinates $x = -2$ and $x = 1$.

$\left[16\frac{1}{2} \text{ square units}\right]$

18 Sketch the curve $y = x^3-2x^2-3x$ between $x = -2$ and $x = 4$. Determine the area enclosed by the curve and the x-axis.

$\left[24\frac{1}{3} \text{ square units}\right]$

19 A vehicle has a velocity $v = (3+4t)$ m/s after $t$ seconds. How far does it travel in the first 3 seconds? Determine also the distance travelled in the fourth second.

(Distance travelled $= \int_{t_1}^{t_2} v\, dt$)

[27 m; 17 m]

20 The force $F$ newtons acting on a body at a distance $x$ metres from a fixed point is given by $F = 2x+3x^2$. Determine the work done when the body moves from the position where $x = 1$ m to that where $x = 4$ m.

(Work done $= \int_{x_1}^{x_2} F\, dx$).

[78 Nm]

# Index

Abscissa, 124
Algebra, Boolean, 192
Amplitude, 27
AND-function, 192
Angles, leading and lagging, 27
  of any magnitude, 3
Areas and volumes, 48, 78
Areas, irregular, 58
  of triangles, 13
  under curves, 213
Average values, 59

Boolean algebra, 192
  laws of, 194

Calculus, 204
Cartesian axes, 124
Centre of gravity, 67
Centroids, 67
Combination of waveforms, 28
Completing the square, 167
Cone, 48
Continuous data, 90
Co-ordinates, 124
Cosecant, 1
Cosine, 1
  rule, 13
Cotangent, 1
Cubic equations, 158, 167
Cumulative frequency distribution, 91
Cycles, 144
Cylinder, 48

Deciles, 101
Definite integrals, 212
Dependent events, 115
Derivative, 206
Determination of law, 125, 133
Differential coefficient, 206
Differentiation, 204, 205
Discrete data, 90
Distribution curves, 107

Ellipse, 78, 79
Equations, graphical solution of, 156
  solution of (algebraically), 167
  trigonometric, 39

Expectation, 115
Exponent, 179
Exponential functions, 179
  graphs of, 179
Extrapolation, 125

First moment of area, 67
Frequency, 28
  distribution, 90
    cumulative, 91
Frustums, 48, 78, 81
Functional notation, 204

Gradient, straight line, 124
  curve, 204
Graphs,
  exponential functions, 179
  logarithmic scales, 144
  solution of equations, 156
  straight line, 124, 133
  trigonometric functions, 4
Grouped data, 90, 100

Histogram, 91
Hyperbolic logarithms, 181

Identities, trigonometric, 39
Inclined plane, 13
Indefinite integral, 212
Independent events, 115
Integration, 212
Interpolation, 125
Irregular areas and volumes, 58

Lagging angles, 27
Lamina, 67
Leading angles, 27
Leibniz notation, 206
Logarithms, Naperian, 181
Log-Linear graph paper, 144
Log-log graph paper, 144

Maximum values, 156, 206
Mean value, 59, 99
Median, 99
Mid-ordinate rule, 59, 219

Minimum value, 156, 206
Mode, 99

Naperian logarithms, 179, 181
Natural laws of growth and decay, 179
Natural logarithms, 181
Normal distribution curve, 107
   probability paper, 108
NOT-function, 192

Ogive, 91
OR-function, 192
Ordinate, 124

Pappus theorem, 79, 85
Parabola, 156
Percentile, 101
Period, 26
Periodic function, 26
   time, 28
Phasor, 28
   resolution of, 29
Planimeter, 58
Prismoidal rule, 78, 83
Probability, 115
   laws of, 116
   paper, 108
Pyramid, 48

Quadratic equations, 156, 157, 167
   formula, 167
   graphs, 156
Quartiles, 101

Rectangular axes, 124
Resolution of phasors, 29
Root of an equation, 167

Secant, 1
Simpson's rule, 59, 219
Simultaneous equations, 156, 157
Sine, 1
   rule, 13
Solution of equations, graphically, 156
   of equations, 167
   of triangles, 13
Sphere, 48, 78, 81
Straight line graphs, 124
Standard deviation, 100
Surface areas of regular solid, 48

Tally diagram, 92
Tangent, 1
Trapezoidal rule, 58, 219
Triangle, solution of, 13
   areas of, 13
Trigonometric equations, 39
   functions, graphs of, 4, 25
   identities, 39
   ratios, 1
Truth tables, 192
Turning points, 156, 207

Volumes of irregular solids, 59
   regular solids, 48

Zone of a sphere, 78, 81

$y = x^2$

$y = 1x^2$

$\dfrac{dy}{dx} = anx^{n-1}$

$= 2 \times 8$

$= 16$